THE RELUCTANT PROPHETS: HAS SCIENCE FOUND GOD?

THE RELUCTANT PROPHETS: HAS SCIENCE FOUND GOD?

By

Stephen Blaha, Ph.D.

ISBN: 0-75966-304-1

This book is printed on acid free paper.

1stBooks – rev. 10/1/01

ABOUT THE BOOK

The Reluctant Prophets: Has Science Found God? explores startling archaeological evidence that shows the concept of God as the Word is at least 5,000 years old and originated in Ancient Egypt. In addition, it shows that the latest scientific theories of the fundamental nature of the universe are linguistic. In these theories the fundamental particles of nature are "letters" in a cosmic alphabet. The universe is thus one tremendous word. The convergence of the religious concept of God as the Word and the scientific concept of the universe as a word suggest that science may be encountering circumstantial evidence of God. The second part of the book considers spiritual phenomena from a scientific viewpoint: haunted houses, miracles, visions, Near Death Experiences and so on. It explores these forms of spiritual experience from a scientific viewpoint. It also considers the evolution of God and discusses God's infinite nature - particularly the type of God's infinity - from a scientific perspective. The second part then inquires into the nature of language, reality and human consciousness. The book is aimed at a popular audience. Concepts are explained in a simple clear way.

To Margaret

and

Natasha, Stephen and John

CONTENTS

PREFACE

Religion tells us God is the "Word". Until now Science has declined to say anything about God: "God is outside the realm of Science's activity. Science is concerned with the material Universe."

What if evidence of God appeared in the makeup of the Universe? Then Science could not excuse itself and plead religion is outside its domain. Instead the study of the universe might reveal aspects of God. We would return to an earlier view: "The heavens reveal the handiwork of God."

There is now some evidence - circumstantial evidence - that Science is at last approaching the fringes of the spiritual - in particular - perhaps the nature of God.

A new view of current theories of the universe has been proposed that may bring science and religion closer together. This new view is not a new theory but instead a new way of looking at current theories. So it does not conflict with existing theories of the universe in any way.

This theory provides a new foundation for "grand unified field theories". It is based on what could be called a quantum computer language. In this language the particles of nature studied by physics (such as electrons and quarks) are merely symbols or "letters" in a vast cosmic alphabet. When they come together in different combinations, they form "words" or the concrete realities that make up the cosmos. These words combine to produce one vast word - the Universe.

In a sense the Gospel of St. John got it right: "In the Beginning [really] was the Word."

This book describes the evidence for the non-technical reader. The technically oriented reader can read *Cosmos and Consciousness* by this author - it contains a detailed technical discussion of the scientific parts.

The book is divided into two parts. The first part shows religious thought for at least 5,000 years - since Ancient Egypt - has supported the concept of God as the Word. Then it shows how scientific thought has evolved to a similar fundamental view of the universe as a word.

The second part of the book describes related topics on spirituality and science. It represents an attempt to apply modern scientific thinking to spiritual issues such as visions, miracles, Near Death Experiences, the evolution of God, the analysis of God's nature (in particular God's infinity), the nature of reality, human consciousness and the relationship between science and religion.

ACKNOWLEDGEMENT

I am grateful to my wife Margaret for reading the manuscript and making many helpful suggestions. I also am indebted to Ann Maruszak for a careful reading of the manuscript and for several very helpful comments. Lastly I would like to thank John Randall, Natasha Randall and Mark Duncan for reading and commenting on the manuscript.

Part I Word - God & Universe

1

An Ancient Concept of God

I have brought the ways of eternity to the twilight of the morning. - Horus

The Concept of One God

Ancient Egypt is the morning of Western civilization. We are more familiar with the gods of Greece and Rome. We are more aware of the philosophy of Greece. But western culture was born in Ancient Egypt.

Ancient Egyptian thought and religion is little known in modern western culture. What is known often

3

sounds alien to modern western thinking. Even in ancient times Egypt was considered remote and mysterious—set apart from other nations.

The philosophic and religious thinking of the Ancient Egyptians was intricate and deep. This depth and range of thought was the result of over a thousand years of relatively peaceful development in organized religious centers that was more or less undisturbed by outside influences or invasions. Egypt was alone in its early civilization. Outside of Egypt there was only primitive cultures and barbarism. The people of Ancient Egypt felt the special uniqueness of their civilization so strongly that they reacted with horror at the thought of a journey outside Egypt.

Egyptian culture developed in splendid isolation. Egyptians developed mathematics, engineering, astronomy and philosophy. Their philosophy sought answers to the eternal questions of life and death. In the opinion of Greek philosophers the philosophy of Egypt was superior to the philosophy of the Greeks. After all Plato and Pythagoras went to Egypt to learn.

The religion of Ancient Egypt was far deeper than other contemporary religions. The Greeks who visited Egypt commented on the depth of religious feeling in the Egyptians and described them as "the most religious of nations." Their concept of the judgment of each individual soul after death – the "weighing" of the heart – is strikingly similar to the modern religious concept of the judgment of each soul after death.

Superficially the Ancient Egyptian religion appears to be polytheistic. We have all seen pictures of their colorful gods that are often half-animal and half-human in appearance.

However the Ancient Egyptian religion was actually monotheistic beneath a layer of polytheism. We see the lesser gods of Egypt pictured in tomb and temple wall drawings. These gods were mortal - capable of dying – in fact, we read of Egyptians hunting and killing them. The myth of the death and resurrection of Osiris is a familiar example. These "mortal" gods do not concern us.

We are interested in the hidden God - the unknown High God of Egypt - the God who was described in the *Book of the Dead*. The concept of the High God was distinctly different from the lesser gods. He did not have the clothing of materiality.

His nature was abstract and spiritual in the modern sense. Even his description in the *Book of the Dead* had a very different style compared to the other parts of the text. Interestingly, this abstract and unusual passage is the passage seen most often in tombs and temples. The

5

description of the High God in chapter 17 of the *Book of the Dead* appears to be the first popular abstract religious text[1]:

'The Word came into being,

All things were mine when I was alone.

I was Rê in [all] his first manifestations:

I was the great one who came into being of himself,

who created all his names as the Companies of the [lesser] gods,

He who is irresistible among the gods.'

Its resemblance to the opening line of the Gospel of St. John "In the Beginning was the Word" is noteworthy.

The *Book of the Dead* is part of the ancient Pyramid and Coffin Texts. These writings, which are the oldest intelligible writings of mankind, stand out for their high quality, and the depth and sophistication of their content. For example, the lamentations for the dead are similar in

---

[1] R. T. Rundle Clark, *Myth and Symbol in Ancient Egypt* (Thames and Hudson Inc., New York, 1993) pp. 78-79.

style and literary content with passages in the Bible and other later religious writing.

The Ancient Egyptians subscribed to the idea of a unique God above all others who created and guided the universe.

An Abstract Concept of God

From its origins, the idea of one God has flowered into the major Western religions: Christianity, Judaism, and Islam. A major concept in these religions, which was also part of the Egyptian concept of one God, is the idea of God as the 'Word'. This concept is extremely unusual compared to the concepts of gods prevalent in Mesopotamia, the Middle East and the Mediterranean region in the period 3,000 BC to 1,600 BC. The normal view of a god was of a being that resembled a man, an animal, a monster or a force of nature. A god was usually represented as an idol of metal and stone or a wall sculpture or drawing.

The abstract view of God as the Word represents a deeper more sophisticated view. It is difficult to understand how this amazing yet strange concept arose. Magic often uses words - casting spells is a common part of magic. Religious ceremonies use words - prayers and song. But it is a leap of imagination - an inspiration - to describe God as the Word. This concept is an extremely abstract characterization of God. Yet it appeared at the dawn of civilization when most cultures viewed gods as animals, forces of nature or people with great powers but human-like emotions.

We will examine this unique concept of God as the 'Word' in the next few chapters and clearly show that this religious concept of God dates from at least 2,500 BC. Then we will see how science, which is based on a totally different mindset, has developed theories of the universe that support a new view of the universe as a 'word'.

This amazing similarity of views - totally unforeseeable until the last few years - raises the very serious question of whether our Science has reached a sufficient depth to detect what may be called a 'spiritual' aspect to nature. In any case this similarity of views can only be viewed as startling.

2

An Egyptian Concept of God

Let there be song and dancing! He has been initiated into the language of the universal gods; he is a uniquely learned one!

Coffin Texts[2], Number 312

Many gods—But One God

Ancient Egypt viewed the gods differently from modern Western man. There was a High God that created the universe and that ultimately governs it. There was also an

---

[2] *ibid.*, p. 146.

assembly of lesser gods that lived and died, and that could experience immortality. But these gods were closer to mortals. They walked the earth, could mingle with Egyptians - and most fantastically of all they could be hunted by the Egyptians. There are even passages describing a hunt for gods in which a god was caught and eaten by the Egyptians. These gods played an important role in the Egyptian view of the universe. But they were closer in nature to angels or demons from the viewpoint of moderns.

The Egyptians also had a high god that closely resembles our modern concept of God. The High God was the ultimate source of all authority. The subordinate 'gods' ruled in his name. The High God was not directly involved in earthly events. Instead, He was the source of power. In the beginning there was a unity - a Oneness of All. But the High God ascended to the sky and separated Himself from creation. God and creation became distinct.

Chapter 42 of the *Book of the Dead* describes the High God[3]:

I am he who is constantly appearing, whose real nature is unknown,

---

[3] *ibid.*, p. 256

> I am yesterday; "He who has seen a million years" is one name of mine,
>
> I pass along the ways of those sky-beings who determine destinies,
>
> I am the master of eternity, ordering how I am fated, like the Great Beetle.[4]

The High God had several names that are more familiar to the modern age: Ra—the High God of the Sun and Atum—'the Complete One'. Like the God of Jews and Christians the High God of Egypt was complete in himself.

The Power of Language

The transition from a High God—mysterious, remote and the source of all power—to the 'Word' may be based on a deep reverence and respect in Ancient Egypt for the power

---

[4] The 'Great Beetle' is the paramount representation of the High God in the New Kingdom's funerary literature. The 'Great Beetle' represents the creative power of God. Beetles emerge - as if by magic - from lifeless waste. (The New Kingdom of Egypt lasted from approximately 1580 BC to 1320 BC.)

of language. To the Egyptian the written and spoken word contained great power to influence events and the world. In part this opinion reflected a belief in magic. But more importantly it reflected a deep reverence for the divinity embodied in language.

Egyptian hieroglyphs had divine power in them. They were more than a form of communication. They were alive and real, and had the ability to cause, and affect, events. As Christian Jacq[5] explains it:

> The Egyptians were so convinced of the power of the hieroglyphs that, in some texts, they cut lions and snakes in half to keep them from causing harm, or held reptiles [in the hieroglyphs] to the ground with knives.

This point is emphasized by Fowden who points out[6]:

---

[5] Christian Jacq, *Fascinating Hieroglyphics*, (Sterling Publishing Co., Inc., New York, 1996) p. 19.

[6] Garth Fowden, *The Egyptian Hermes*, (Princeton University Press, Princeton, NJ, 1986) pp 63-64.

> The ancient Egyptians believed, not only that an object's or being's whole nature was implicit in its name, so that knowledge of the name conveyed power over what it designated, but also ... that a supernatural force was inherent in the actual written or engraved letters that made up the name.

Words were the source of power over the universe. The chants of priests, the spells of the magician and the writings on the temple walls influenced the course of events. From this mindset it is natural to attribute great power to the names of the gods and to attribute ultimate power to the name of the High God. The last step—which the Ancient Egyptians took—was to identify the High God with the Word—a profound leap of logic.

God is the Word

The Egyptians clearly and unequivocally called the High God the Word. They also said the universe and all things in it were created by the words uttered by the High God. The universe and its contents are the embodied words of God.

The concept of God as the Word is strikingly brought out in the Egyptian *Book of the Dead*—the oldest written document in the world[7]:

> I am the Eternal, I am Ra ... I am that which created the Word ... I am the Word...

Or in an alternate translation[8] of Spell 307 of the *Book of the Dead*:

> I am the Eternal Spirit,
>
> I am the sun that rose from the Primeval Waters.
>
> My soul is God; I am the creator of the Word.
>
> Evil is my abomination, I see not.
>
> I am the creator of the Order wherein I live,

---

[7] P. Barguet, *Le Livre des Morts*, (Ed. Du Cert, 1967) p. 123.

[8] Clark, *ibid.*, p. 77.

> I am the Word, which will never be annihilated
>
> In this my name of 'Soul'

Another of the many references in Ancient Egyptian literature to God as the Word[9]:

> My lips are the Twin Companies; I am the great Word...

The concept of the High God of Egypt as the 'Word' was widespread in the literature of Ancient Egypt and a central theme of the deeper parts of Egyptian religion.

The Wonder of God as the Word

How did the Ancient Egyptians arrive at such a sophisticated view of God? We do not see this view in other contemporary civilizations (3,000 BC - 1,000 BC) in Palestine (excepting Israel), Babylon or Sumer, or further East in India or China. Nor does it appear in Greece or Asia Minor.

---

[9] Clark, *ibid.*, p. 60.

It appears that it started with the love of the written word of the Ancient Egyptians. The belief in the power of words leads the association of words with God and, through a leap of thought, to the identification of God with the 'Word'. This leap of thought is not automatic. Most cultures have a belief in magic - the power of words - but Ancient Egypt was the first to make the leap to the concept of God as the Word. Afterwards we see this concept in Judaism, Christianity and Islam.

Were the Ancient Egyptians Divinely Inspired?

The appearance of the 'Word' in Ancient Egypt raises an issue for Christians and Jews who regard it as a special part of their religious traditions. How can we understand the existence of the 'Word' in the religion of Ancient Egypt? Who told them?

Christian belief accepts the idea that pagans can acquire knowledge of the one true God through reason or enlightenment from God. The Christian sees his knowledge of God as coming not only from reason but also, and more importantly, from revealed truth.

Judaism also accepts the fact that non-Jewish religions and peoples can acquire a knowledge of, and a relation to, God. Perhaps the best example of this is the non-Jewish Canaanite priest-king Melchisedech to whom Abraham paid a tithe.

3

A Jewish Concept of God

Abraham's father was a maker of idols. Abraham had an encounter with God in Haran that led to the Covenant. Subsequently, the Hebrews had many contacts with Egyptian culture. Egyptian armies crossed Palestine several times to war in Syria in the period from 2000 BC to 1000 BC. Egyptian trade extended to Lebanon. Hebrews visited Egypt as well including a stay in Egypt that lasted over four hundred years.

There is significant evidence that the Hebrews adopted ideas and religious practices from the Egyptians.

This is understandable considering the long history of contact between these peoples and the high civilization of the Egyptians. The Greeks, for example, considered Egyptian philosophy, religion and culture as deeper than their own.

Remote God and Living God

The Egyptians had many gods that lived and "walked the earth." And they had a High God that was remote and yet was the source of all power.

Judaism, on the other hand, is a monotheistic religion - one God. However there is a very old part of Judaism represented by the Kabbalah, and other groups, that has a somewhat different view of the Divinity. (It must be emphasized there is a diversity of views of God in Judaism.) In the Kabbalistic view there is a remote God (called *Ein Sof*) and a 'living' God. The remote God created the universe and then retired to a distant role. "And on the seventh day God ended his work which he had made; and he rested on the seventh day from all his work which he had made."

The 'living' God, who may be viewed as an alternate facet of the Deity, tends to the day-to-day affairs of the universe. This Living God is the God that supervises the affairs of men.

The Jewish view of the remote God and the living God is similar to the Ancient Egyptian view of a High God and lesser gods (that attended to earthly concerns). Of course the Jewish view is uniformly monotheistic. There is only one Living God. However in Abraham's time, and until at least the time of Moses, the Hebrews acknowledged the existence of other gods.

The Hebrews chose to follow the God of the Covenant. But they had lapses into idolatry that are recorded in the Bible. In addition the Ten Commandments acknowledged other gods in the wording of the commandment "thou shalt not have other gods before Me". This commandment implicitly recognizes the existence of other gods. The prophets did not accept these other gods as true gods.

A major difference between the Hebrews and the Egyptians is that the Egyptians worshipped their lesser gods while the Hebrews were commanded to ignore other gods.

Language, Reality, God's Name

Judaism has always had a deep reverence for the Hebrew language and for its Holy Scriptures.

The Hebrew language, like hieroglyphics for the Ancient Egyptians, has often been viewed as having magical effects - particularly by the Kabbalists. Words had power and could affect the world. Unlike the Egyptians where the pictorial aspects of a hieroglyph had significance (a snake hieroglyph might lead to a snakebite) the effect of Hebrew words and passages was based on their numeric value. The letters of the Hebrew alphabet were numbered starting with א (aleph) having the value 1.

The association of numbers with the letters of the Hebrew alphabet, and the calculation of the value of words, was called *gematria*. The Kabbalist calculated the numeric value of words. Words and passages with the same numeric value had a related significance. A description of a typical Kabbalistic approach is[10]:

> In the writings of Eleazer of Worms, for example, are found tracts on magic and the effectiveness of God's secret names as

---

[10] Dan Cohn-Sherbok and Lavinia Cohn-Sherbok, *Jewish and Christian Mysticism*, (Continuum Publishing, New York, 1994) p. 36.

well as recipes for creating the golem (an artificial man) through letter manipulation. Another feature of this movement concerned prayer mysticism. In the literature of the pietists attention was given to techniques for mystical speculation based on the calculation of the words in prayers, benedictions, and hymns. The number of words in a prayer as well as the numerical value were linked to biblical passages of equal numerical value as well as designations of God and angels. Here prominence was given to the techniques of *gematria* (the calculation of the numerical value of Hebrew words and the search for connections and other words of equal value)

The Jewish Kabbalists viewed language as Reality. The Kabbalistic concept of reality is language-based and centered on the Name of God. Kabbalists use language to obtain knowledge of nature and God's reality. The idea of the reality of Language is especially evident in the thinking of Abraham Abulafia, a Thirteenth Century mystic whose theories were widely adopted in Kabbalist circles.

Abulafia developed a form of contemplation based on the Name of God. The method consisted of contemplating combinations of Hebrew letters that could be viewed as components of the Name of God. The process of contemplation brought one into a mystical union with God.

The Name of God plays a central role in Abulafia's theory and in Jewish mysticism in general. God's Name is viewed as an entity in itself. The Name of God embodies the content of the universe. Everything derives its meaning from it.

Abulafia's theory was based on the Kabbalistic concept that reality is composed of sacred language. The universe, and everything within it, is language in its innermost reality. Our common view of the universe as material in nature is an illusion.

Abulafia thought that a special language existed that reflects and expresses the thoughts of God. The letters of this language were both the components of the deepest spiritual reality as well as the components of the profoundest level of physical reality.[11]

The idea of language as the central reality of nature is extremely interesting when compared to Egyptian

---

[11] See Gershom Scholem, *Major Trends in Jewish Mysticism*, (Schocken Books, New York, 1995).

thinking. Clearly the Kabbalistic view of reality is remarkably similar to the Egyptian concept of the central role of language in the creation and maintenance of the universe. For both Ancient Egyptians and Kabbalists language is sacred and language is reality. The universe is one vast hieroglyphic.

Another important Kabbalistic figure who performed gematria as part of meditative practices is Isaac Luria, a leading Sixteenth Century mystic. Cohn-Sherbok[12] points out:

> In Lurianic teachings there are various [kabbalistic] meditative procedures related to specific actions. Such meditation brings an individual into the upper spheres and involves the combinations of divine names. The essence of Luria's meditative system consists of unifications (*Yihudim*) in which manipulations of the letters of the name of God take.place.

---

[12] Dan Cohn-Sherbok and Lavinia Cohn-Sherbok, *Jewish and Christian Mysticism*, (Continuum Publishing, New York, 1994) p. 53.

The Baal Shem Tov, founder of the modern Hasidic movement, added to this thought by stating[13]:

> Whenever you offer your prayers and whenever you study, have the intention of unifying a divine name in every word and every utterance of your lips, for there are worlds, souls, and divinity in every letter. These ascend to become united one with the other and then the letters are combined in order to form a word so that there is complete unification with the divine.

The Word is the Name of God

One might object to identifying the concept of the Name of God with the concept of God as the Word. However, these concepts are very closely related if not identical in the final analysis. An argument in favor of this identification begins with the observation that the Name of God has enormous power in the eyes of Kabbalists. (Consider also how the Orthodox Jews do not speak or

---

[13] *ibid.*, p. 64.

write of G_d directly out of reverence and fear.) If God's Name has such power then it must derive that power from God. It's power must be part of God's power. If this is so, then the Name of God must be part of God as well. Since God in reality can have no parts - God is one undivided whole - then the Name of God is in some sense God. Thus the Word.

Words Created the Universe

Just as the words of God created the universe and all it contains according to Ancient Egyptian thought, the words of God created the universe in the Biblical account in Genesis. "And God said let there be light ... " There are enough comparable descriptions in Ancient Egyptian writings and in Genesis to suggest an Egyptian influence in Genesis.

For example, in the Ancient Egyptian account of the creation of the universe God caused the emergence of the land (the Primeval Mound) from the Primeval Waters. In Genesis God says "Let the waters under the heaven be gathered together unto one place, and let the dry land appear." The concept of the universe in Genesis begins with a sky filled with waters. Beneath it is a firmament

(Heaven). Beneath the firmament were the lower waters. God created land amongst the lower waters.

The Egyptian view of the universe was similar. The universe was composed of water in all directions. The sky separated the waters from the waters and played the role of the firmament. The Primeval Mound appeared in the lower waters below the sky.

Both the Hebrews and the Egyptians viewed the universe as a "bubble" in the midst of the universal waters.

Conclusion

At this point it is clear that there are remarkable similarities between Ancient Egyptian and certain aspects of Jewish thought. These similarities are summarized in the following statements:

1. Words have power to influence and change reality.

2. The substance of Reality is a divine language - words - at its deepest level.

3. God's Name - the Word - is the source of power and the deepest meaning and totality of existence.

4. God's words created the universe (Genesis).

4

A Christian Concept of God

In the beginning was the Word, for the Word ... was the eternal beginning, ... a revelation of the Eternal One.[14]

Jakob Boehme

The Christian views of God are diverse. All Christians share a belief in one God and in the New Testament as well as the Old Testament. Perhaps the best known description of God in the New Testament is the

---

[14] *ibid.* p. 140

beginning of the Gospel of Saint John, "*In the Beginning was the Word, and the Word was with God, and the Word was God.*" This surprising expression of faith places the Word – language – at the core of the reality of the Universe and, in fact, makes it an attribute of God Himself. The Word is with God and the Word is God.

Later in the First Chapter of St. John he says, "And the Word was made flesh, and dwelt among us..." describing Jesus as the embodiment of the Word.

From where did John's abstract view of God come? The depiction of God as the Word has no obvious connections with other parts of the New Testament or the Old Testament.

There are two possibilities. One can view these passages as an inspiration of St. John or as reflecting the thinking of a contemporary group in Israel or the Middle East at the time St. John wrote his Gospel.

St. John was a Jew familiar with current ideas in Judaism, and at the same time a leading proponent of Christianity. He probably would have had some familiarity with the group that became the Kabbalist sect. The Kabbalists had a similar reverence for the Name of God.

The Kabbalists became a prominent part of Jewish culture and thought in the Middle Ages. But according to Jewish written and oral tradition, their roots extend back to the First Century – the time of St. John. Whether St. John was influenced by the First Century precursors of the Kabbalah or other sources is not relevant for our discussion. What is relevant is the central role of the Word - the Name of God.

For both the Kabbalah and St. John (as well as other Christian saints and mystics) the Word was the fundamental form of reality.

The Christian Word

Mystics in the Christian mystical tradition showed a continuing use of the Word as a metaphor or reference for God. This thread was probably at least partly the result of Kabbalistic influences that may have started as early as the First Century AD as evidenced by the beginning of the Gospel of St. John.

In the Third Century the famous theologian Origen[15] "engaged in Scriptural exegesis. In his opinion, the Incarnate Word is implicit in the Hebrew scriptures".

The famous Christian mystic Jakob Boehme[16] said:

---

[15] *ibid,.* p. 8.

[16] *ibid.,* p. 140.

"The beginning of all being was the Word, as God's breath-forth, and from Eternity God has been the Eternal One and remains thus in eternity. But the Word is also the flowing out of the divine will or the divine knowledge. For just as thoughts flow out of the mind and the mind still remains a unity, so also was the Eternal One present in the outflow of the will. It says: In the beginning was the Word, for the Word as the flowing-out from the will of God, was the eternal beginning, and remains so eternally, for it is a revelation of the Eternal One, by which and through which divine power is brought into a knowing of something. (We understand) by Word the revelation of the will of God, and by God, we understand the hidden God, as the Eternal One, out of which the Word springs eternally."

The tone of this passage is remarkably similar to passages in the Egyptian *Book of the Dead.*

Another similarity to Ancient Egyptian thinking appears in a poem by the Seventeenth Century "prophet of the ineffable", Johannes Scheffler[17],

"God has all names and none

Indeed one can name God by all his highest names

And then again one can each one withdraw again."

In the Nineteenth Century Elizabeth Catez wrote a prayer containing[18]

"Oh eternal Word , Word of my God, I want to spend my life in listening to you. … create in my soul a kind of incarnation of the Word: that I may be another humanity for him in which he can renew his whole mystery."

---

[17] *ibid,* p. 141.

[18] *ibid,* p. 147.

Marie of the Incarnation, a Seventeenth Century Christian mystic described a vision[19] in a passage that ended with the sentence:

> "The Eternal Father was my father, the adorable Word, my spouse, and the Holy Spirit was he who by his action worked in my soul, fashioning it to support the divine impressions."

And in the Twentieth Century Karl Rahner, the noted theologian, viewed Christ is the embodiment of the mystical Word.

Thus there is a steady progression of Christian religious-mystical thought from the First Century AD until the Twentieth Century focused on God (and Christ) as the Word.

The Power of Words

For many Christians the most important example of the power of words is the transformation of bread and wine into the Body and Blood of Christ in the Mass. This alone

---

[19] *ibid,* p. 137.

shows the power of words for a major part of Christianity: Catholics, Anglicans and Lutherans.

There is also a Christian tradition of magic and alchemy that shows a belief in the power of words to affect changes in the world. The best known example of this part of Christianity is the legend of Faustus. Faustus first becomes involved in magic, then makes a pact with the devil and then uses spells together with Mephistofeles to influence people and events. Words - spoken and written (the pact with the devil) - are a crucial part of the legend.

Words Created the Universe

In accepting the Old Testament Christians accept Genesis with its account of the creation of the universe through the words of God. So Christians share with the Ancient Egyptians and Judaism a belief in creation through the words of God.

Conclusion

At this point it is clear that there are remarkable similarities between Ancient Egyptian, Judaic and Christian thought on the Word. It seems that these similarities reflect a historic flow of concepts from Ancient Egypt to Judaism to Christianity. In addition there is probably a direct influence of Ancient Egyptian thought as understood in the First and Second Centuries AD on the Christian church in Egypt.

Egypt had a profound influence on Christianity. The best example of this influence is the origin of Christian monasticism in the deserts of Egypt. There are also more subtle influences on Christian Theology and Philosophy - particularly from Alexandria - the melting pot of Egyptian and Greek thought. One is reminded here of a similar modern blending of traditions that has taken place in Brazil and South America between Spanish Catholic, Indian and African elements.

The similarities of Ancient Egyptian, Judaic and Christian that we have seen are summarized in the following statements:

1. Words have power to influence and change reality.

2. God is the Word.

3. God's words created the universe.

5

God as the Word

The Thread of the Discussion

In the preceding chapters we have seen a continuous thread of thought starting in Ancient Egypt at least five thousand years ago and extending to the present day: God is the Word and the universe was created with words. These concepts are not scientific - they are the result of religious thought - and, in the opinion of believers, the result of divine inspiration.

In this chapter we want to appreciate the uniqueness, originality and continuity of these concepts. In succeeding chapters we want to examine a new scientific view that has remarkable similarities to these concepts. If these similarities hold up under close scrutiny we can hope that we are beginning to reach a depth of physical theories that may directly address issues and aspects relating to God - at least through His interaction with the natural world. Then we may be able to say that Science has at last found evidence of God.

Uniqueness and Originality of the Concept

If we survey the religions of the world we see that only Egypt (the oldest civilization) developed the concept of God as the Word. One can relate the origin of this concept with the deep reverence that they had for writing and the spoken word. Many ancient writers have commented on the beauty and the inspirational qualities of the chants and prayers in ancient Egyptian temple ceremonies.

The Ancient Egyptians attributed strong magical powers to the hieroglyphic writing. The extension of the power of writing to the power of God and then the identification of God as the Word seems a natural progression. As natural as it seems the civilizations of Sumeria, India and China did not develop this concept.

Points of Contact between the Spiritual and the Material Realms

If the concept of God is the Word and the concept that the universe is a word have any relation to each other then we have physical evidence that there is a relation between the spiritual world and the material world. In fact the long existing religious belief that the material universe is a reflection of God would now have a concrete foundation. The fact that the universe is a word reflects the fact that God is the Word.

While science deals with the material universe and religion deals with the spiritual realm there has to be some points of contact between them. If there were no points of contact then we would have no knowledge of the spiritual realm since we are obviously material beings - at least in part. There are three evident possibilities:

1. Man (intelligent life?) has both a spiritual and material part. Man (intelligent life?) is the point of intersection between these realms.
2. The spiritual can affect the material realm directly.
3. Both of the preceding possibilities.

Western religions support the third possibility. Western religions believe that the material realm can affect the spiritual realm through prayer or magic, and the spiritual realm can affect the material realm.

Western science doesn't affirm or deny any of these points. Human consciousness is so complex that it is difficult to begin to understand the material parts of consciousness let alone the spiritual parts which are undoubtedly more subtle. Thought transference, ESP, and other quasi-spiritual phenomena have not been susceptible to scientific analysis.

The scientific analysis of spiritual effects in the material realm - miracles and so on - has also not been successful. Phenomena are too complex and usually susceptible to differing interpretations. Spiritual events are also not normally reproducible - they happen once. Science, on the other hand, deals with reproducible phenomena.

The net result is that a scientific proof of the existence of the spiritual is a very difficult task.

6

The Universe as the Scientific Word

New Hope for Contact - The Ultra-Small

There is a new possibility for a scientifically verifiable point of contact with the spiritual - the ultra-small. Science has delved deeper and deeper into the nature of the material world - the atom - sub-atomic particles - the nature of space and time itself. As a result science is beginning to understand the very fabric of creation. Questions are being raised by physicists about hidden dimensions, space and time warps, time travel and so on.

43

These topics have been stock in trade for spiritual speculations for hundreds of years.

So one can hope that we may be able to see indications of points of contact with the spiritual realm at this deep level.

Concept of the Universal Word

Surprisingly, there is a new view of science in the ultra-small that is based on language and leads to the statement that the universe can be viewed as one vast word. In this view the particles of matter at the sub-atomic level are letters in a cosmic alphabet.

If the universe can be viewed as composed of particles that are letters does it not constitute a gigantic word – a word that is continually evolving into the future?

This picture evokes several possibilities. One possibility is the universe is God's utterance of his own Name. Another possibility is the universe is in some sense part of God – is God. A third possibility is both of the previous possibilities.

All of these possibilities reflect an amazing similarity between the five-thousand-year-old concept of God as the Word and the new view of the universe as a word.

The material universe that we think of as space, time and matter is in reality a "word" uttered from the beginning of the universe and extending to the end of time.

The next few chapters build a picture of the material world, and of computer languages that lead to the linguistic view of the universe.

7

Levels of Physical Reality

Sub-Atomic World

The material universe, as we know it, has a hierarchical structure with many levels: the universe itself, super-galaxies, galaxies, star systems, planets, chemical compounds, atoms, and the sub-atomic world. We live at a level of matter that consists of chemical compounds. The earth and its creatures are composed of these compounds.

After about 500 years of what may be called modern science we have developed an understanding of the sub-atomic world of matter. Most of that understanding has

47

come in the last seventy-five years—starting with the development of quantum physics.

The modern picture of matter starts with the statement that matter is composed of atoms. Inside the atoms are particles called electrons that circle a central nucleus just as the planets circle the sun. An atom consists mostly of empty space. The electrons occupy very little space in an atom. The nucleus is also very small compared to the size of the atom.

The nucleus of an atom is composed of still smaller particles called nucleons. There are two types of nucleons: protons or neutrons. Protons have a positive electric charge and neutrons have no electric charge.

Quarks, Electrons and Other Particles

Each nucleon is, in turn, composed of another kind of particle called a quark. Quarks are wonderful things. Even their name is whimsical coming as it does from *Finnegan's Wake* by James Joyce. No one has ever "seen" one in isolation. Yet the majority of knowledgeable physicists are convinced they exist. Whether quarks are composed of yet another kind of particle is not known.

At this point we have a picture of ordinary everyday matter consisting of layers – things composed of things within things within things. The lowest known level is composed of quarks and electrons (and other kinds of particles). There are eighteen varieties of quarks. The building blocks of normal matter, protons and neutrons, are made of quarks in differing combinations.

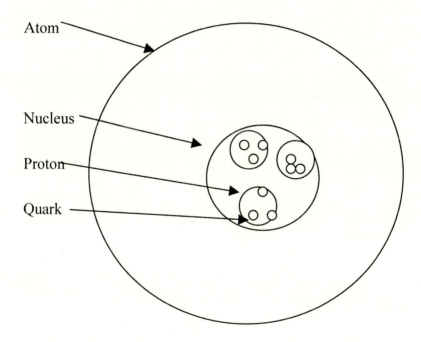

Figure. A depiction of the particles in an atom. The figure is not drawn to scale. The atom is much, much bigger than the nucleus.

Electrons are as fundamental as quarks. They are elementary particles in their own right. There are thirty-eight plus kinds of elementary particles: quarks, electrons, muons, neutrinos, gluons, W bosons, photons and so on. These particles form the "cosmic alphabet."

The detailed nature of the fundamental particles of nature is an interesting story. But our focus is on their role as letters or symbols within a cosmic alphabet. The next chapter carries forward the theme from particles to the current theories about their nature. Following that discussion we see these theories lead to a view of particles as letters in a language.

8

Physics Theories of Reality

Current Theories

There are a number of theories of the fundamental nature of our universe. The most established theory is called the Standard Model of Elementary Particles. It appears to be solidly supported by experiments. The other theories are part of a family of theories called SuperString theories. Many physicists believe SuperString theories are a deeper, more fundamental approach to the understanding of the universe. However experimental evidence for

SuperString theories is extremely skimpy at this point in time 2001.

In this chapter we will briefly look at fundamental theories and see how they all lead to the concept that the universe is a word.

The Standard Model

The Standard Model .is a theory that directly describes the elementary particles found in nature: quarks, electrons, muons and so on.

It appears to furnish an almost complete theory of elementary particles describing both the particles and almost all the forces that they experience: electromagnetism, the weak interactions and the strong interactions. The gravitational force between particles is not part of the theory.

The Standard Model has a classification scheme for particles built into it that is reflected in the following tables:

The Standard Model Fermion Family - Matter

Generation	Quarks				Leptons
I	Up	u_1 u_2 u_3	μ_e	electron neutrino
	Down	d_1 d_2 d_3	e	electron
II	Charmed	c_1 c_2 c_3	μ_λ	muon neutrino
	Strange	s_1 s_2 s_3	λ	muon
III	Top	t_1 t_2 t_3	μ_τ	tau neutrino
	Bottom	b_1 b_2 b_3	τ	tau

The Standard Model Gauge Fields ("Force Fields")

Force	Number and Symbols of Particles
Weak & Electromagnetic	**1** W_0 Name: W boson
Weak & Electromagnetic	**3** W_i $\quad$ i = 1,2,3 Names: W bosons
Strong	**8** G_i i = 1,2, ..., 8

	Names: gluons

The Standard Model also contains a particle called the Higgs particle.

The classification of particles in the preceding tables is the result of over seventy-five years of experimental and theoretical analysis. The tables are analogous to the periodic table of the elements in Chemistry. The detailed form of these classification tables is an interesting subject covered by many popular books on modern physics (to which the interested reader is recommended).

The Standard Model incorporates almost all of our current experimental knowledge of particles and interactions (except gravity) into a single theory.

While the Standard Model is not in clear disagreement with any current experimental data there are a number of problems that the Standard Model does not appear to resolve:

1. Explaining the "dark matter" found in galaxies
2. Incorporating gravity into the Standard Model
3. The appearance of "too many" arbitrary constants in the Standard Model theory

4. The lack of "elegance" in the Standard Model
5. The "weird" pattern of symmetries in the Standard Model

Despite these problems the Standard Model is an extremely successful theory and a triumph for the physicists – experimental and theoretical – whose labors over the last seventy years have led to its creation.

The Standard Model is a form of theory called a quantum field theory. This is an important fact. It leads to a view of matter where the elementary particles listed in the preceding tables are "letters" or symbols. These "letters" can combine to make words (matter) with the ultimate word combination being the universe.

SuperString Theories

The best current attempts to develop a deeper type of theory than the Standard Model are called SuperString Theories. These theories attempt to explain the nature of elementary particles and the forces between them based on the assumption that a particle is a kind of mathematical string vibrating in a space of many dimensions.

Particles in the Standard Model are point-like. SuperString Theory treats each "fundamental" particle –

quarks, electrons, and so on – as an extended string in space-time. The strings are so small that their string-like nature cannot be verified experimentally with today's experiments. Current experiments see particles as point-like. If experiments could probe more closely they might see particles as strings. Strings are thought to be approximately 100 billion billion times smaller than a proton.

A string in a SuperString Theory is a continuous, one-dimensional space (like a thread) representing one particle. It can be visualized either as a rubber band (closed string representation of a particle) or as a piece of cotton thread with two ends (open string representation of a particle).

●

Figure. Particles are point-like in the Standard Model.

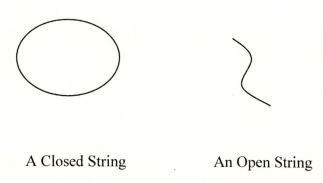

A Closed String   An Open String

Figure. In SuperString Theory particles are strings.

In SuperString Theories each elementary particle is a particular vibration of a string. A string can vibrate in a number of different ways just like a violin string can vibrate at different frequencies. Each elementary particle corresponds to a particular type of vibration.

In twenty-five hundred years we have moved from the music of the spheres of the Greek philosophers to the music of Twentieth Century SuperStrings: from divine harmony to SuperString melodies.

9

Pure Creation

Particles are Discrete

The ancient Greeks pointed out that matter could be continuous or that matter could be composed of discrete particles. Modern Physics tells us that matter is composed of discrete particles. There are thirty-eight currently known fundamental particles.

The existence of discrete particles is an important fact. It provides a basis for our fundamental theories of matter and the forces of nature. It is particularly important when we consider the creation and annihilation of particles. In the Nineteenth Century matter was thought to be everlasting and indestructible. In the Twentieth Century we came to realize that particles could be created and destroyed. Matter was no longer indestructible. Einstein caused this change with his famous $E = mc^2$.

When particles are created or destroyed they are created or destroyed in integer quantities. We cannot create (or destroy) half a particle or three-quarters of a particle. This fact is embedded within our theories of elementary particles: the Standard Model and SuperString theories.

The creation (or destruction) of whole numbers of particles is a basic feature of a general type of theory called a quantum field theory. The Standard Model is a quantum field theory. SuperString theories are similar in this respect to quantum field theories. So all the known theories of elementary particles share this feature of creation or destruction in whole numbers.

Particle Creation and Destruction

Modern physics has developed the means to understand particle creation and destruction. The first steps in this direction took place at the beginning of the Twentieth Century when physicists first realized that light is composed of particles called photons. Photons are commonly emitted (created) and absorbed (destroyed) in everyday life. The basic processes involving light and electromagnetism are the emission and absorption of photons by electrons. These processes are normally pictured using diagrams like:

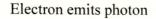

Electron emits photon   Electron absorbs photon

Figure. Photon emission and absorption by electrons. The solid lines are electrons and the dashed lines are photons - particles of light.

These diagrams are called Feynman diagrams after the physicist who first used them - Richard Feynman.

Photons can also annihilate (or be destroyed) by changing into an electron - positron pair. This process represents the creation of matter from light. A positron is the anti-particle of an electron. It is in a sense the "opposite" of an electron.

61

Another process involving matter transformations is the annihilation of a positron and an electron to produce a photon that carries off the energy that was previously in these particles (matter - antimatter annihilation). This process creates light from the destruction of matter. We describe this process as the creation of energy from the annihilation of matter.

The following diagrams illustrate these processes:

   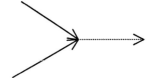

A photon annihilates into an electron and a positron

An electron and a positron annihilate into a photon

Figure. Creation and annihilation of electron-positron pairs. The solid lines are electrons and positrons (anti-electrons). The dashed lines are photons.

There are many processes for particle annihilation and creation in the Standard Model. But in all cases particles are created or destroyed in whole number quantities. You can't create or destroy part of an elementary particle. The same situation is true in SuperString theories.

Particles are often created in high-energy collisions at particle accelerators. If we cause two particles to collide at high energies new particles can be created. We can picture the collision process as:

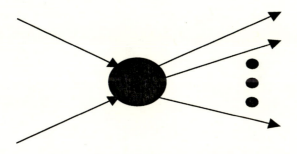

Figure. Two particles collide and create additional particles. The dark blob in the middle represents the interaction of the incoming particles that results in the production of the output particles.

As far as we know the Standard Model of elementary particles successfully accounts for particle creation and annihilation as we see it in experiments. SuperString theories also may be able to describe these processes.

The creation and annihilation of particles in elementary particle processes appear to be the only real acts of physical creation and annihilation taking place in our universe.

Data Creation in Computers

While particle creation and annihilation may be the only real physical creation and annihilation processes, there is another phenomenon in which creation and annihilation takes place: in data transformations within a computer.

A simple example of data transformation in a computer is to imagine a computer that takes every key we press on the keyboard and displays it on the computer screen twice. For example in the below picture we type ABC on the keyboard and see AABBCC appear on the computer screen.

Figure. A PC with input typed on the keyboard and output displayed on the screen.

We call the letters typed on the keyboard the *input* and the letters appearing on the computer screen the *output*. We see in this example a form of pure creation—the creation of data in this case. We can create a diagram for this process of data transformation (creation) that looks like the diagrams for particle creation:

Figure. A diagram for a data transformation in a computer.

In a sense a computer creates (destroys) data characters just like interactions can create (or destroy) particles. So we can see an analogy between the transformations of data characters in a computer, and particle annihilation and creation. *Nothing else in nature is so directly analogous to particle creation and annihilation.*

10

Reality as Language

The Standard Model Defines a Language

The close analogy between particle creation and destruction, and data transformations in a computer is deeper than the discussion of the previous chapter would suggest. An examination of the Standard Model Theory of elementary particles from a linguistic viewpoint[20] shows that this theory actually defines a type of computer language.

---

[20] Stephen Blaha, *Cosmos and Consciousness*, (1stBooks Publishers, Bloomington, Indiana, 2000)

A language consists of letters (symbols), words and a grammar. The letters in the Standard Model language are the elementary particles of nature: quarks, electrons, photons and so on. The words are particle combinations (states). The grammar is specified by the Standard Model theory.

A grammar states how letters and words can combine and transform from an input set of words to an output set of words. This is similar to the grammar of a language such as English where the grammar rules specify how letters and words combine to form sentences.

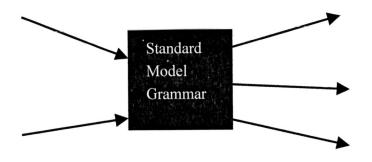

Figure. Example of a particle interaction: An input word – two particles (symbols or letters) collide and the Standard Model grammar produces an output word consisting of three particles (symbols or letters) with a certain probability.

There is one complication in the Standard Model language. The language is inherently quantum probabilistic. An input set of particles (letters or words) can generally result in many different possible output letters or words. Every possible output has a certain probability of occurring. These probabilities are calculated according to the rules of quantum physics.

However, despite these complications, the point remains that the Standard Model defines a language where the letters are particles.

The Surprisingly Simple Standard Model

Many physicists feel that there are too many elementary particles in the Standard Model. They say that thirty-eight or so particles are too many. Well, from the viewpoint of our linguistic approach thirty-eight "letters" is not much different than the number of letters in many human languages such as English, Hebrew, Russian and so on.

From the point of view of a language theorist *the simple finite language representation of the Standard Model is a very special situation.* As Hopcroft and

Ullman[21], noted computer linguists, point out, "there are many more languages than finite representations." Languages can have infinite alphabets or infinite grammars or other complications.

From this point of view our universe is very special. *The physical equivalent of an infinite alphabet would be a universe with an infinite number of different particles of matter. Every particle of matter could have a different mass as well as differing in other properties. From this point of view our universe and the Standard Model are truly marvels of simplicity.*

SuperString Theories are also Linguistic

SuperString theories are also linguistic in nature.[22] They have to be linguistic because they must account for the same particle picture that the Standard Model does with

---

[21] J. E. Hopcroft and J. D. Ullman, *Formal Languages and Their Relation to Automata*, (Addison-Wesley, Reading, MA, 1975) page 2.

[22] See reference 20. It shows that the fourier expansion of a SuperString in terms of raising and lowering operators can be reinterpreted as a representation of a quantum computer. The result is a detailed demonstration of the equivalence of SuperString theory and quantum computers. Quantum computers have a language and grammar.

particles as "letters", and with particles being created and destroyed just like data in a computer.

In a SuperString Theory the "letters" are the SuperString particles. We think of each particle in a SuperString theory as some vibration of a closed string or open string. So each letter represents a specific SuperString vibration.

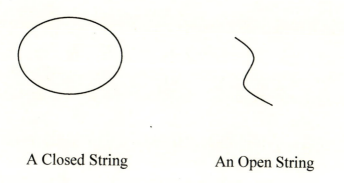

A Closed String　　　　　An Open String

Figure. In a SuperString Theory the particles are strings.

In an earlier section we saw how a computer can take input data such as ABC and transform it into output data such as AABBCC. Similarly SuperString particles can undergo transformations with one SuperString particle becoming two SuperString particles, or two SuperString particles combining to make one SuperString particle.

We can specify this symbolically with rules like:

Rule 1.　　　◎→　　◎ ◎

Rule 2.　　　◎◎→　◎

where the symbol ◎ represents a SuperString particle. These rules look vaguely like chemical reactions. But each of them is actually a form of grammar rule for the linguistic view of SuperString theory. The set of SuperString "letters" plus these grammar rules define a language just like a human language is defined by a set of letters - an alphabet - and a grammar.

We can picture Rule 1 as a SuperString breaking into two SuperStrings:

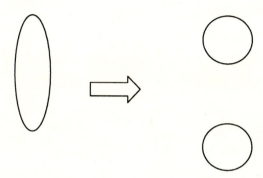

Figure. One diagram representing Rule 1 for the "fission" of a SuperString into two SuperStrings.

.

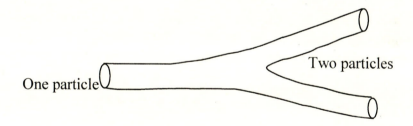

Figure. Another visualization of Rule 1 for the "fission" of a SuperString into two SuperStrings.

Rule 2 can also be easily visualized with similar pictures.

Thus we see that SuperString theories also can be viewed as linguistic. They lead to a picture of the universe as one vast word.

11

The Universe as a Word

Language as the Essence of the Universe

The five hundred years of modern scientific development leading to the Standard Model of Elementary Particles can be viewed as five hundred years spent in the discovery of a language of nature – the language embodied in the Standard Model (and in SuperString theory).

Yet a critic might say "it is all very well that you can show that the current fundamental theories of the universe can be viewed as defining a language, but there is also matter and space and time in the universe. These things are real. The universe is more than language."

A closer scrutiny of space-time and matter suggests that the solid fabric of reality may not be as substantial as we might believe based on our experience of everyday experience. First we note matter itself is composed of fundamental particles. Second we must realize that space and time dissolves on the most fundamental level into gravitons - particles of gravity. So we come to the view that everything in the universe is somehow a particle or a combination of particles.

Next we look more closely at matter. Matter also is not quite what we think when we look at it closely at the sub-atomic level. We think matter is solid. But when we look into matter at smaller and smaller distance scales the solidity disappears. Imagine a super-microscope that peers into a lump of matter constantly increasing the magnification. First we see that atoms are almost entirely empty - just a few electrons circling a very small nucleus containing protons and neutrons.

Then we look into the protons and neutrons only to find quarks. This brings us to the current level in the investigation of the fundamental nature of matter. Matter is mostly empty except for these "small" fundamental particles.

Quarks, electrons and other particles appear to be fundamental in the sense that they are not known to be composed of "smaller" particles.

The Nothingness of Matter

This new view that we are presenting differs greatly from our everyday perceptions of solid tangible matter. How could reality be so different? Well, the solidity that we think we perceive is actually the effect of electromagnetic force fields.

The "matter" in a solid object only occupies a minute fraction of the volume of the object. As we look more and more deeply into matter the proportion of empty space grows.

Electromagnetic fields keep us from putting our hands right through a lump of matter. The electromagnetic fields of our hands repel the electromagnetic fields in matter and prevent our hands from penetrating the matter. If we were not affected by electromagnetic fields we could put our hands through material objects.

Have We Reached the Deepest Level of Matter?

Let us assume for the moment that we have reached the level of truly fundamental particles that are not composed of yet more fundamental particles. In this case we have either the Standard Model or a SuperString theory as the only valid candidates for the theory that correctly describes them.

If the Standard Model is the correct theory of matter then matter is not very substantial. An elementary particle has certain formal properties such as spin and charge. But the feature that we most attribute to matter, namely, mass ("weight") actually is the result of an interaction with another particle called a Higgs particle. Basically a particle is nothing in itself. A particle is really a structure or order placed on "nothing". All the aspects of matter that we normally think of - mass, energy, charge and so on can be viewed as structural in origin, and not "real" in the usual way we understand reality in everyday life.

If a SuperString theory is the correct theory of matter the same observations apply. Particles are defined as structural things and, in themselves, are not substantial or real from the viewpoint of everyday life. In these theories a particle is a mathematical string - not a substantial thing in our view.

So we must conclude that the only thing that exists at the most fundamental level is structure in the form of a language with particles being letters in that language. Language is the only reality.

What If There are Deeper Levels of Matter

It is possible that our current understanding of elementary particles will change and a new level of matter will be discovered. Perhaps quarks, electrons and so on are made of yet more fundamental particles. This possibility is not ruled out experimentally.

Perhaps there may be many more fundamental levels of matter. At each level the so-called fundamental particles may be composed of still more fundamental particles.

Suppose this progression reaches an end at some level. At that level we would then find a theory of the Standard Model type or of the SuperString type. (These types of theories are the only ones that we know of that have a particle interpretation with particle creation and annihilation.) This situation would again lead us to the view that the universe is structure and the structure is linguistic.

If there is no end to the levels of matter does that not squeeze matter down to nothing as the levels unfold into "smaller and smaller particles." Here again matter degenerates into structure and the structure is linguistic at each level since we would expect the approximate theory at each level to be a quantum field theory with a linguistic interpretation.

Thus we are led in all cases to a linguistic universe. The sum total of the particles in the universe constitute a word.

12

Has Science Found God?

Starting at least 5,000 years ago man has had an image of God as the Divine Word. We do not know how this inspired concept arose - whether it was through Divine inspiration or through a logical development of ideas resulting from hundreds of years of thought and meditation. But the facts are clear that it is a central idea of the Ancient Egyptian religion, of Judaism, and of Christianity.

Today, the deepest theories of science have been found to have a linguistic representation. The fundamental particles of nature can be viewed as letters in a cosmic alphabet. And the universe can be viewed as a vast word.

The startling convergence of the scientific view of the universe as a word and the long held religious view of God as the Word is amazing from any perspective. The words of St. John, "In the Beginning was the Word" now have a dual interpretation as a statement about God and a statement about the Big Bang - the origin of the universe.

The final chapters in this story remain to be written. The growth and deepening of our religious concept of God will continue. Now it may start to go hand in hand with our deepening knowledge of science. We are not after all anywhere near *the last syllable of recorded time - just at the beginning of the Word.*

Part II Spirit & Science

13

Spiritual vs. Material

Eminent scientists such as Stephen Hawking have said there is no evidence for God. They state that wherever we look we see no need to invoke a deity to explain what we see.

Another plausible view is that the universe is set up to work perfectly well according to physical laws instituted by God. God "built it right" in the Beginning and so He does not need to intervene continuously or even very frequently.

Does He intervene? The question of evidence for spiritual events is a difficult one. We often hear of strange events that are interpreted as spiritual or mystical. How can we tell if an event is "beyond science" and therefore a spiritual manifestation? The standard "scientific" thinking that we see in movies, novels, and pop science programs, is to place all sorts of complex scientific instruments at the place of an event – say a "haunted house". Somehow the scientific instruments will reveal whether the event is spiritual or material (explainable by science).

Actually scientific instruments only provide data that must somehow be interpreted. For example, suppose we detect "ghosts" in a haunted house with scientific instruments. Are these "ghosts" spiritual? The question is not easy to answer. The detected entities could be material beings from "hidden" dimensions. (Speculations on hidden dimensions are currently popular in physics journals.) So we would have to test whether the detected phenomenon was a material phenomenon due to hidden dimensions.

It would be a major undertaking to positively prove that such "ghosts" were not material entities.

Probably a giant particle accelerator would have to be installed in the "haunted" house (leaving very little of the house remaining) and subtle measurements made of space-time effects on particle interactions to test for evidence of other dimensions. (Another problem: putting equipment in the house might eliminate the phenomenon.) If other dimensions were not detected, then their existence could still not be ruled out ("too subtle to detect").

Furthermore, the experiments would require the cooperation of the ghosts. If the ghosts did not reliably reappear then the scientific analysis would not be possible. Ghosts are not normally considered patient and cooperative!

The net result is that proving an event is spiritual is very difficult. Ruling out all possible material causes is not easy.

A Procedure for Determining An Event is Spiritual

The foregoing discussion raises the question: Can we develop a general procedure to distinguish spiritual events from material events? To answer this question let us consider a specific event that might be spiritual and might be material: Sacheverell Sitwell (a brother of Edith Sitwell) wrote a book on vampires approximately one hundred years ago. One event he described in his book was that of a tomb in Italy containing a young girl in a glass coffin. Although the girl had been dead for hundreds of years she looked as she did in life and her blood had not congealed. Sitwell suggested that she might be an example of vampirism or "life" after death thus raising the question of a spiritual effect at work.

A Spiritual Phenomenon?

How do we know whether this event is spiritual in nature or a strange material effect? If the body is examined chemically in detail and no explanation is found for the preservation of the body then there are two possible conclusions. The preservation is of spiritual origin or the preservation is explainable by scientific analysis but we don't yet know enough to perform this analysis (perhaps we will in the future). The result of this thought process is that our analysis will result in a decision for a scientific explanation or it will be inconclusive.

A Material Phenomenon?

Science is, in a sense, predatory. It grows by explaining new phenomena. When science encounters a phenomenon it responds in one of three ways: 1) it understands the phenomenon based on current knowledge, 2) it extends current knowledge to develop an understanding of the phenomenon (thus extending the domain of science), or 3) it says it cannot currently understand the phenomenon.

If science cannot understand a phenomenon then there are two possibilities: 1) the phenomenon is reproducible in which case it becomes a mystery of science for a future generation of scientists to explain or 2) the phenomenon is not reproducible - sometimes it happens; sometimes it doesn't. In this case science will tend to dismiss it as an illusion or as unverifiable.

Science is Favored in this Material World

Science cannot PROVE a phenomenon is spiritual. Science can show a phenomenon is scientifically understandable and thus material.

Because science is growing, and does not claim to know everything, a reproducible, but not currently understandable, phenomenon can only be classed as a material phenomenon for the future to explain.

Since we think a spiritual event is normally caused by the action of a spirit it is unlikely that a spiritual event will be reproducible. We cannot expect a spiritual entity to perform an action on demand if the same experimental conditions are set up again. Angels do not devote themselves to making scientists happy with reproducible spiritual phenomena.

So if a spiritual entity causes an event that is scientifically measurable then the event would probably be classified as a fluke since it would not be reproducible.

Random inexplicable events are fairly frequently encountered in many areas of science such as cosmic ray physics and high-energy physics. These events are usually discarded if they do not occur again in subsequent experiments. Spiritual events are therefore apparently always going to be outside the scientific sphere.

Any phenomenon that science can explain either with present knowledge, or an extension of present knowledge, or in the future through new knowledge is material and not spiritual.

Are Flying Saucers Real?

The idea that we are being visited by beings from other worlds is at least 150 years old. Charles Fort described incidents in which people saw spacecraft such as rocket ships in the Nineteenth Century. Since roughly 1948 there have been a stream of stories about incidents involving flying saucers.

People claim to have proof of the existence of alien spacecraft in the form of eyewitness reports, radar sightings and photographs. Based on our previous discussion on proving the reality of spiritual events scientifically it is clear that it would be equally difficult to prove the existence of flying saucers and aliens using the type of evidence that has been gathered over the years. There is always an alternate conventional explanation of this type of evidence. So the only way to prove the existence of aliens, flying saucers and so on is for a prolonged event during which multiple scientific tests could be performed to verify its extraterrestrial origin.

Is Matter non-Material?

If the matter in the universe consists of small, indivisible particles as the Greeks (Democritus) originally envisioned then we could say we lived in a material universe that was very different from the non-material, spiritual realm. The material universe would then be distinctively different from spiritual entities.

But the universe has proven to be very different from our everyday concept of matter when we examine its fundamental structure. The solid veneer of matter has dissolved into a phantasmagoric ghost-like structure governed by a theory with a linguistic appearance.

The distinction between the spiritual and the linguistic view of the universe is not so clear cut. A purely spiritual entity has form but does not have material substance. Can we say an electron, as described in modern theories, is much different? It has structure - form - but it does not have substance. It acquires its mass ("weight") through its structure - in particular, through its structural relationship with Higgs particles. (Higgs particles are special particles that give mass ("weight") to all the massive elementary particles. At this writing the first experimental evidence of the existence of Higgs particles is appearing.)

Is an electron, in itself, spiritual? An electron intrinsically has no mass in the Standard Model and no discernible material features. It only has form.

A purely spiritual entity seems to be something that has no material parts and yet exists. In addition spiritual entities are expected to somehow have the attributes of individuality and will. A spiritual thing does something. It might worship God; it might aid or hurt humankind; it may be undergoing torments for past crimes. An electron fails to be spiritual for at least these reasons: lack of individuality and lack of will. On the other hand it could be viewed as fundamentally non-material (in everyday language). The same reasoning applies to the other elementary particles found in nature.

So we conclude that the universe is made up of non-material things - the fundamental elementary particles. The universe can thus be viewed as nothing more than a vast word.

Visions - Spiritual or Scientific?

Man has experienced visions since ancient times. The Bible and other ancient sources record many visions. A famous vision is the vision of Ezechiel of the "wheel within a wheel". This vision has been interpreted both as a spiritual vision and as a (material) sighting of an alien space ship.

Many normal modern people have claimed to see visions. In some cases such as the vision of the Lady of Fatima large groups have seen a vision.

What are we to think of visions? Clearly, if we "see" a vision we either perceive it through normal vision (meaning it was conveyed to the eye through rays of light) or we perceive the vision through our mind's "eye" internally in our consciousness. In the second case there is no material vision in the usual sense. Also the interpretation of an internal vision is more difficult: is it a psychological event or is it a mental event caused by an external agency - perhaps a spirit?

External World Visions

Let us consider the first case of a vision perceived as an external event conveyed to the eyes by light rays. The light rays are part of the material universe. Is the vision spiritual or material? The answer is in the source of the vision. If the light rays originate from a known material source due to a known physical process then the event is material. There is still a possibility of a spiritual agency causing the material event to happen in a way consistent with physical laws. The miracle at Fatima might illustrate this dilemma. Assuming it is not a case of mass hallucination there was a peculiar optical phenomenon involving the sun in the heavens. The phenomenon could be due to unusual atmospheric conditions. The fact that it happened at the precise time predicted for a miracle might be coincidental. Scientifically one could account for the optical phenomenon. But the timing of the event - it didn't happen the day before and evidently it has not been seen since - gives a believer sufficient grounds to call it a miracle.

This discussion leads us to conclude that external visions conveyed to the eye via light may be entirely material in origin (in this case the scientist ought to able to reproduce it in the laboratory) or may appear to be material but in fact may be spiritual in origin. (A spiritual agency guides material events in a way consistent with physical law to create a miracle.)

There is also the possibility of a vision that has a recognizable form (perhaps of a religious figure) and that lasts sufficiently long (perhaps days) and that is apparent to so many individuals so as to remove all doubt as to its spiritual nature. This type of vision has not been seen in modern times.

Internal Mental Visions

There appears to be a type of vision that we perceive through our mind's "eye" within our consciousness. This type of vision is not manifestly material. Is it a psychological event? Or is it a mental event caused by an external agency - perhaps a spirit?

We could try to make a distinction between visions seen when awake and visions seen while asleep. However it is not clear that this is really a meaningful distinction. We tend to attribute more significance to a vision seen while awake because a vision occurring while asleep may be difficult to distinguish from a dream.

Probably the best starting point for understanding internal visions is to view a vision as a particular mental state (or series of mental states) of human consciousness. Whether there is a spiritual part of human consciousness or not it seems reasonable to think that a vision somehow is embedded in the physical human brain. Science has shown the emotions, memory, visualization, thought and the sense of self ("I am me as opposed to you.") are all physical phenomena with the complex of cells that constitute the brain. Since a vision normally involves most if not all of these aspects of consciousness the vision must be viewed as physically manifested in the brain.

The big question is who creates the vision. Experience with mind-altering drugs has shown that artificial visions can be created chemically. Experience with meditation and mental training has shown that visions can be stimulated and perhaps created (a widespread belief of Buddhists and others) using these techniques. Therefore we must conclude that visions can be created through natural or material means.

Suppose a vision were created through spiritual means by God or by a spirit. How could we distinguish an internal vision of this type from a vision generated naturally through a progression of psychological states? One possible way of differentiating between natural visions and spiritual visions lies in the content. If a person experiencing a vision "sees" things that are totally foreign to their experience, then one can reasonably think that the person has received input from an external (spiritual?) source.

But there are a number of possible external sources including information from a prior life (reincarnation) and ESP to name a few. The strange content of a vision could also be a creative combination of a person's prior experience. So the significance of the content of the vision is not clear-cut.

There does not appear to be a factor or set of factors that could distinguish between a vision generated spiritually and a vision generated through material means. Besides the content of a vision the only other significant factor normally is the timing of the vision. Visions often occur "at the right time". For example a woman, AMM[23], lost a friend in 1990 who was like a mother to her and with whom she had forged a deep bond of affection over a period of twenty years. Within hours after her friend passed

---

[23] This poignant experience was directly recounted to me by the person involved who we will call AMM.

away she experienced a vision of the friend in which she saw a momentary, happy, golden image of her friend looking down from a position near the ceiling. Her departed friend's natural daughter who was sitting with her did not see the vision because it was fleeting. So it must probably be viewed as AMM's internal vision. In AMM's words, "I instantly experienced an overwhelming feeling that our "mother" went straight to heaven and was now experiencing eternal love and joy. ... That event was one of the greatest moments of my life and changed me forever."

This example shows the close timing of the vision with a major event in the life of the visionary. It also shows a great, and long lasting, depth of emotion and feeling resulting from the vision.

The vision could be viewed as a true vision generated by the spirit of the departed friend. AMM chose this interpretation. But it could also be viewed as part of the psychological grieving process. In this second view we then have to ask whether the vision was generated naturally or through spiritual means. From the nature of the event and in the current state of our knowledge of the human mind we cannot answer this question.

But there is an issue here that we can consider. Can the human mind be influenced by spiritual means in a way that does not conflict with the natural behavior of the mind?

As we learn more about the human mind we increasingly see that consciousness is based on physical brain activity. Because the human brain contains enormous numbers of cells that are connected in incredibly complex ways the behavior of the brain must have aspects that are statistical (probabilistic) in nature.[24] The statistical aspects introduce an element of chance.

The brain does not proceed from state to state in an ironclad deterministic way like a conventional computer. Chance intervenes. For this reason it is possible for a spiritual entity to guide the development of states of the brain in such a way as to generate mental phenomena such as visions without explicitly violating physical laws. With this approach natural laws are followed but spiritual effects can be implemented that do not contravene these laws.

On one occasion Einstein said, "The Lord is subtle but he is not malicious." and on another occasion he said, "God does not play dice with the universe." In this situation we might say, *"So subtle is the Lord that He occasionally uses slightly loaded dice."*

Spiritual effects can take place in a manner that is consistent with physical law by taking advantage of

---

[24] See *Cosmos and Consciousness* by this author for more details.

99

statistical probability on the classical level of everyday experience and quantum probability on the subatomic level.

Internal visions can be spiritual in nature and yet not conflict with scientific laws of nature.

Cause or Effect?

Can divine intervention be understood within the framework of science? For example, can God instantly cure a desperately sick child in a way that our medicine will find understandable? Or must it be an unexplainable miracle?

One possible way of having miracles within a scientifically acceptable framework is based on a scenario due to Richard Feynman. This scenario allows "scheduled" divine intervention through a form of "predestination". Feynman made the important observation that physical laws are often simpler and more convenient if one assumes the future as well as the past determines events.

We normally think that the present evolves from the past. Feynman raised the possibility that the present may depend on the future as well as the past. It is possible according to accepted laws of Physics that the present is also determined in part by the future. In this case, we are, in a sense, locked in by both the past and the future.

100

This view is similar to Calvinist Predestination. It does not necessarily eliminate free will so it is not strictly Calvinist. (Calvinism is a Christian sect that was started by John Calvin in Switzerland during the Protestant Reformation. Calvin believed that God determined the future completely so man did not have free will.)

Using the Feynman conjecture we can understand that God, knowing both past and future, can "set up" the past so that a miracle in the present happens in harmony with physical laws. Thus Hawking's comment about the absence of evidence for God in nature merely shows that God is acting more subtly than we thought.

The Reluctant Prophet: Has Science Found God?

14

The Analysis of Near Death Experiences

Nature of Near Death Experiences

Near Death Experiences (NDEs) raise interesting questions about the nature of consciousness and its relation to the spiritual. In a NDE a person enters a state similar to death with all life signs disappearing. The person is clinically dead: no brain function or other vital signs. After revival the person may have memories of events happening while the person was clinically dead - an NDE.

One recent study of sixty-three heart attack patients who were successfully revived found that seven of them had a memory of the time in which they were clinically dead while the remaining fifty-six had no memory of the period.[25] Four of the seven with memories had vivid memories of activity: moving around, talking to others, and thinking while their brains were measured to be not functioning. These four patients were identified as having an NDE.

The NDEs of these patients consisted of a variety of memories including heightened senses; speeded up time; feelings of harmony, peace and joy; seeing a bright light; loss of bodily awareness, entering another realm of being; communicating with dead relatives; and so on. One person had an encounter with a mystical being.

One possible cause of an NDE - low oxygen levels - was not present in the case of these patients.

In a follow up study Dr. Parnia found 3,500 individuals who had apparently experienced an NDE - vivid memories of the time in which they were clinically dead. These individuals had a variety of experiences including the above-mentioned experiences and experiences such as out of body experiences in which they

---

[25] Dr. S. Parnia et al, article in the medical journal, *Resuscitation*, February, 2001.

felt they were outside their body and looking (often down) at it and the surrounding scene.

The incidents that Dr. Parnia and his colleagues investigated as well as many other NDEs reported over a period of centuries raise profound questions of life and death, and of consciousness and its relation to the brain. Is an NDE a spiritual experience or a physical experience? Is an NDE a phenomenon of the brain or of consciousness or of both? Is an NDE an indication of life after death? Is the existence of NDEs evidence for a human soul?

Simple Models of Consciousness and the Brain

In order to approach these questions we must have a model or theory of consciousness and the human brain. A model built along scientific lines will enable us to visualize and consider the issues in an organized concrete way. A detailed model is beyond the current state of knowledge. However we can create simple models that provide a framework for discussion.

Actually there are two simple models that can be constructed based on the simplest ideas of the brain and consciousness - ideas that hopefully will continue to be viable as we learn more about consciousness. These models both assume that consciousness only communicates with the brain and the brain communicates with the body as needed.

Let us assume that human consciousness has two parts - a spiritual part and a material part. We assume we can differentiate between the two parts in a simple way: the material part disappears when the human body is totally dead; the spiritual part is the part that remains when the human body is totally dead. (Whether the spiritual part is the soul or part of the soul is an issue we will not address since there is no experimental data to resolve it.)

The material part of consciousness is assumed to be directly connected to the brain and, practically speaking, indistinguishable from it. Modern brain research has uncovered numerous direct connections between specific brain areas and specific features of consciousness. If the brain area is not operational (i.e. not working) then the corresponding feature of consciousness is absent.

With these considerations in mind we can create two models specifying the relations of the brain and consciousness. These models assume the body is connected to the brain (as it is in fact) and that consciousness is connected to the body solely through the brain.

The difference between the models is the communication channel between the hypothesized spiritual part of consciousness and the body. In model I (see figure following) communication between the spiritual part of consciousness and the brain take place through the material part of consciousness.

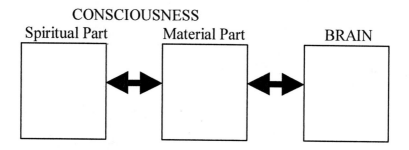

CONSCIOUSNESS

Spiritual Part   Material Part   BRAIN

Figure. Model I in which the spiritual part of consciousness communicates with the brain through the material part of consciousness. The arrows indicate communications channels. We assume communications proceed in both directions between each of the components.

If an NDE occurs and Model I is the correct model any sensory impressions must proceed from the brain through the material part to the spiritual part.

Spiritual events (such as "seeing" a mystical being) must proceed through the material part in the reverse direction to become embedded in the brain. Model I avoids the issue of a mismatch between the material part of consciousness and the brain. You can't change the brain without first changing the material part of consciousness.

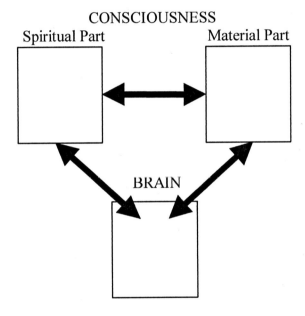

Figure. Model II in which the spiritual part of consciousness communicates with the brain directly. The arrows indicate communications channels. We assume communications proceed in both directions between each of the components.

The second model, model II, allows the brain and the spiritual part of consciousness to communicate directly. Model II (see preceding figure) does have a potential issue. A spiritual event can change the brain while leaving the material part of consciousness unchanged unless the brain automatically updates the material part.

Can We Experimentally Distinguish Between the Brain and Consciousness?

The potential mismatch between the material part of consciousness and the brain can be made more graphic by considering a "thought experiment" (i.e. an experiment carried on in thought not in a laboratory). Suppose we imagine a person starts to do a mathematics calculation in his/her head such as repeatedly doubling starting from 1. So the person calculates: 1, 2, 4, 8, 16, 32, 64, and so on. Then suppose we turn off the area of the brain associated with this type of mathematical calculation. Then after a short interval we turn the brain area on again. Will the person pick up where they left off when the brain area was turned off or will the consciousness have continued to calculate in the interval with the result that there appears to be a leap in the calculated numbers when the brain area resumes activity? A version of this experiment might be performed with animal subjects.

If there is no calculation done while the brain area was turned off then the brain appears to be linked to the material part of consciousness. If the material part of consciousness cannot do anything without support from the brain then it must be viewed as part of the brain and generated from the brain as we assumed when constructing our models.

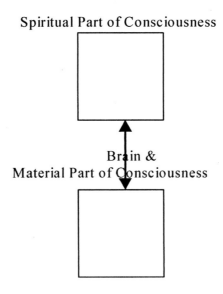

Figure. Model III based on identifying the brain and the material part of consciousness.

The general drift of modern brain research suggests that the material part of consciousness cannot perform activities such as mathematics without the direct participation of the brain. So we identify the material part of consciousness with the brain.

After identifying the material part of consciousness with the brain we see that models I and II become model III (see preceding figure). Model III allows us to analyze certain aspects of NDEs. As we discussed earlier an NDE typically has a visual part in which the individual may see the scene surrounding his/her body and may converse with individuals.

If model III is correct and if an NDE is a purely spiritual phenomena (no material part) then several questions arise. How does the spiritual part of consciousness perceive a visual scene? After all, "seeing" a recognizable group of individuals gathered around the patient in an NDE requires the ability to detect visible light of the normal frequencies that the human eye sees. But in an NDE the human eye supposedly is effectively turned off - almost no brain function. Can the spiritual part of consciousness directly see? It seems unlikely. And how would it choose to see in precisely the same frequencies as the eye? Why wouldn't it see in other frequencies and thus perceive a much different scene? The scene would look very different if seen in the infrared frequencies or x-ray frequencies or radio wave frequencies. Lastly, how does a purely spiritual entity see material electromagnetic frequencies?

Based on this type of logical inquiry it seems that an NDE is not a spiritual experience. Somehow an NDE is a material phenomenon - a form of dream or hallucination that is perhaps formed around sensory inputs to the brain that occur at a very low level while the brain function is almost absent.

A View of NDE's

A person may have an NDE after a sudden traumatic event such as a heart attack that may cause an immediate shutdown of the brain. The shutdown is both sudden and almost complete. However a low level of brain activity continues. This level of activity is usually viewed as too low to support organized activities of consciousness.

In contrast a person injured in a car accident may enter a state of unconsciousness less suddenly. The unconscious state is less severe with most brain functions still operative although the person has no memories of the period of unconsciousness. In this state a person does not experience the memories of an NDE.

A comparison of these situations suggest that the speed of the shutdown of brain functions may be the crucial factor in determining whether an NDE can occur. In the shutdown of brain function leading to an NDE the individual may not go through the "normal" shutdown sequence and as a result the senses and memory may remain operational in some sense. (Compare this situation with the improper shutdown of a Personal Computer such as a sudden power failure. When the computer is restarted it has bad "memories" that need to be removed before the computer can function properly.)

The relevant factor in determining whether a transition is sudden (opening the door for an NDE) or not is the reaction time of the brain's chemical response to the traumatic event (heart attack, car accident, ...) If the transition takes place suddenly before the brain can react it may lead to a state with an NDE. If the brain has time to react normally then an ordinary state of unconsciousness can result.

In a sudden transition, in which the brain cannot react normally, the brain may follow an alternate course of action and release natural soporific and mildly hallucinatory chemicals that simulate out of body experiences, serenity, detachment, visions and imaginary conversations with individuals.

After all these same sorts of experiences often occur in dreams while asleep. So the human brain is quite capable of simulating these experiences. In addition NDE-like experiences can be induced using drugs. Shamans use natural drugs to generate NDE-like experiences.

After the person returns to a normal state after an NDE the brain may integrate and weave the sense impressions and other brain events occurring in the NDE into a coherent set of memories. This process can take place in a time interval of the order of a fraction of a second. So the person emerges into a normal state with "solid" memories.

This analysis suggests that an NDE is a natural phenomenon tied to brain chemistry. Should we be disappointed that it does not appear to be a spiritual phenomenon? Not really. After all does God need a person to be near death in order to communicate with him/her?

Some people suggest an NDE is a partial entry into a spiritual state ("approaching the pearly gates of Heaven"). Then the person is "sent back" to an earthly existence and revives. This idea is based on the assumption that God doesn't know whether the person will die or revive, and so begins the person's entry process to Heaven. It appears to conflict with the concept of an all-knowing God. Of course, it is also possible that the person in an NDE may make a decision about whether to die or not using free will. Perhaps they may want to return to life for some good reason and make that decision while in the borderline state between life and death with a serene view of the approach of heaven.

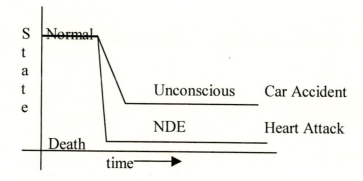

Figure. A suggestive illustration of the change in the state of brain function of a person due to a sudden heart attack vs. due to a car accident where the transition to unconsciousness may be more gradual. A person may have a more sudden decline of brain function in a heart attack setting the stage for an NDE.

States of Consciousness

There appear to be a number of distinct states of consciousness. Actually there is probably an infinite gradation of states. And the distinct states that we identify are probably families of states that are similar and therefore can be logically grouped together. Here is a list of states of consciousness that have been described in the literature and some associated features and points of differentiation.

Normal & Terminated

These states are the normal states of consciousness that occur in life or death.

Ecstatic

It appears that we are capable of entering states of ecstatic consciousness through meditative practices or through drugs. These states can be roughly characterized as states of "low" ecstasy in which the consciousness relaxes into a receptive state or states of "high" ecstasy in which the consciousness enters a state of heightened awareness and response. An example of a possible state of high ecstasy is the state that contemplatives enter when they achieve a form of "union" with the Godhead. The entire being of the contemplative is awakened and the contemplative's body may shake violently.

116

An example of a possible state of low ecstasy is the receptive state that contemplatives enter when they feel that they have merged with the universe or universal consciousness. Buddhist meditation practices seem to lead to this type of state.

Sleep

When a person is asleep consciousness enters a state (or set of states) that is different from the waking state. The bodily functions and mental functions are different from the waking state and from the state of unconsciousness. Sleep states are currently under intense investigation. They are characterized by reduced bodily functions, reduced and different brain activity and an intermediate state between consciousness and unconsciousness characterized by some awareness of the outside world and the body as well as dreaming.

Unconsciousness

The state of unconsciousness is seen reasonably often in everyday life. A common cause is receiving a hard blow on the head. Unconsciousness appears to be reached "slowly" (although we may think it quick) through "natural" bodily processes although it may happen quickly in terms of wall clock time. The slowness of the transition to unconsciousness allows the body to go through a shutdown sequence disconnecting the senses and turning off the brain. The result is no memory of the period of unconsciousness.

Zombie State

It appears that certain chemicals/drugs found in nature can cause a person to enter a state where all bodily and most mental functions are reduced to a minimal level that may give an appearance of death. In this state an individual's senses may still perceive events and the individual's memory may or may not be operational. This state seems to be similar to the state of a person experiencing an NDE. A difference seems to be that a zombie can use its senses such as sight or hearing but it does not apparently have out of body or other NDE-like experiences. It would be interesting to attempt to simulate an NDE-like experience using "zombie drugs" in a controlled scientific experiment.

Vestigial Hibernation/Suspended Animation

When a person enters a state of hibernation (assuming a vestigial form of hibernation exists in man - the Rip van Winkle state) or suspended animation (a possibility that may exist in the near future) then the

118

person's consciousness enters a state that probably is similar to a state of unconsciousness. Mental processes, the higher senses (sight and hearing), and memory are probably not operational. It would be interesting to test whether dreaming is possible in this state. (Current scientific experiments have studied dreaming in sleeping dogs. Similar experiments could be performed on hibernating creatures and man.)

NDE State

An NDE state can be generated by a sudden shutdown of brain functions as a result of a heart attack or some other sudden traumatic event. The individual may not go through the "normal" shutdown sequence and as a result the senses and memory may remain operational in some sense. In addition the brain may release natural soporific and mildly hallucinatory chemicals that simulate out of body experiences, serenity, detachment, visions and conversations with individuals that are similar to shaman experiences that are induced by drugs.

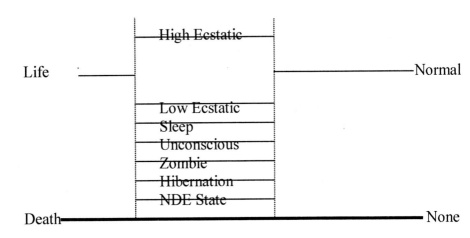

Figure. A visualization of the states of consciousness between life and death. The ordering of the states of consciousness is based on an approximate "level of brain functions" for each state of consciousness. The distances between the levels is not significant in this schematic diagram. The appearance of the High Ecstatic state above normal life is based on the concept of a state wherein life functions - particularly brain functions - are raised above the norm. There are ways of moving towards this higher state such as taking stimulants or eating high caloric foods.

15

The Reluctant Prophets of Science

"Fringe Science"

A scientist would normally prefer to remain in his or her laboratory or library, and do the work of science - the investigation of natural phenomena. However the extreme importance of scientific developments for society has made it impossible for scientists to enjoy their isolated existence. We see the practical importance of scientific discoveries in the technological progress of society. We see the intellectual importance of scientific ideas as they pass

121

from science to philosophy, theology and other branches of learning. The theory of relativity and quantum theory have had a profound impact on philosophy, art, literature and other intellectual pursuits.

Because of the spectacular success of science in recent centuries science has attracted individuals who have only a superficial knowledge of science but who develop concepts and theories - quasi-science theories - and then ask for a hearing from professional scientists. These theories are often called "fringe science" because they are on the fringes, or beyond the fringes, of science as scientists understand it.

Scientists are pleased that people take an interest in science and that they try to play a role in its advancement. On the other hand scientists have precious little time for their work let alone time to discuss theories that they can often tell at a glance are unacceptable. What's worse, the proponents of these theories often become more persistent as a scientist gently tries to put them off.

The net result of this fairly common situation is that scientists tend to develop a thick skin and try to avoid fringe theories and their proponents.

Why Is Science Wary of Religion?

Against this backdrop of a continuing barrage of pseudo-science scientists encounter an interest in science from philosophers, theologians and religiously minded people. The people of religion fall into a number of categories: the idle curious, the people fearful/angry at scientific ideas that conflict with their religious ideas, and the people who accept science and try to integrate it with their religious ideas.

The groups that see a conflict between science and their religious beliefs sometimes respond by calling for the suppression of scientific ideas.

A familiar example is the theory of evolution. The theory of evolution appears to be different from the Biblical account of creation. On the other hand, after well over a hundred years of conflict, science has come up with "Eve" - a being living a few hundred thousand years ago that genetic evidence suggests is the mother of all of humanity. A believer in the Biblical story of creation might take comfort from this scientific fact with an "I told you so".

A believer in Biblical creation could also look at the "Big Bang" scientific theory of the creation of the universe and see that it has similarities to the Biblical account of creation. Perhaps God was describing creation in words that ordinary people of the time could understand.

A reasonable person of a religious turn of mind can accept the revelations of science as what they are - hard won knowledge grounded in fact. If the scientific result is in conflict with their beliefs then the religious person might say, "Keep at it, friends. Your theory will eventually evolve and become consistent with my religious convictions. After all you found Eve a hundred years after Darwin."

This attitude would help enormously in removing the conflicts between science and religion.

Another less obvious level of conflict is generated by the appropriation of scientific ideas by a religious group to support their teachings. For example, introducing quantum or relativistic notions into religion, morality or philosophy can change the labels or nomenclature in these areas. But these scientific ideas do not and cannot prove or demonstrate any principles in these areas - not even by analogy. Attempts by individuals to appropriate scientific ideas and bend them to their own purposes makes scientists generally wary of these individuals.

Guided Science

Perhaps the most difficult issue in the relation of science and religion is the fear of scientists that allowing religious interpretations of their work may lead to attempts by religious leaders to control or constrain science.

Religion generates strong emotions. People will die or kill for religious causes.

Science also has an emotional component. But it is relatively restrained. And scientists generally try to avoid introducing emotion into scientific debates. Scientists do not kill other scientists over differences in scientific ideas.

A very real danger is that scientific ideas may have direct implications on religious issues leading to attempts to control science by religious groups. The controversy over evolution is a perfect example.

The net result is that scientists will generally try to avoid suggesting their work has religious implications.

Despite this attitude some scientists and writers interested in science try to establish connections between modern scientific ideas and religious or philosophic ideas.

Amazing Coincidences between Science and Religious Thought

Many books and articles have been written showing that certain religious or philosophical ideas of ancient vintage appear in this or that scientific theory. In most cases these assertions are correct. Many ancient ideas are similar to recent scientific results in a qualitative way.

What shall we make of these apparent coincidences - if they are coincidences? There appear to be four possible explanations of this type of coincidence:

1. The ancient idea was the result of divine inspiration and it was left to modern science to provide a concrete proof of the idea.
2. The ancient idea was implanted in ancient times by some unknown agent that might be extraterrestrial in origin.
3. The ancient idea was the result of inspired logical thinking or an intuitive leap by an individual or group.
4. The ancient idea was a lucky guess.

It is impossible to decide between these possibilities in an absolute way. It boils down to a matter of belief. So we each make our choice. And we realize that these "coincidences" do not prove anything in a mathematical or scientific sense. But they may constitute a proof within the framework of an individual's mindset.

Why Scientists are Reluctant Prophets

The preceding discussion shows why scientists are reluctant prophets:

1. The twisting of scientific ideas to produce fringe scientific theories that are then peddled to scientists.
2. Attacks on scientific ideas that appear to conflict with religious beliefs.
3. The unwarranted appropriation and extension of scientific ideas that appear to support a religious or philosophic belief.
4. The danger that non-scientists may attempt to control scientific investigation if science impinges on cherished beliefs.
5. The attempt to use similarities or coincidences between scientific thought and religious thought to prove religious beliefs.

All of these possibilities can potentially have disastrous effects on the future of science.

How Should Science and Religion Relate?

For these reasons science and religion should probably be kept separate from each other. The future of science should be in the hands of the scientists as the future of religion is in the hands of religious leaders and God.

However the cross-fertilization of ideas between science and religion can be of value. Science and Religion are both components of an overall world-view. Similarities and differences are worth exploring. Common concepts can be a thought-provoking source of future growth in both disciplines.

There is an overlap between science and religion in the "Big Picture" concepts: the question of the nature of the universe, the question of the origin of the universe, the question of the evolution of the universe, the question of the end of the universe, and the question of the evolution of Man and possibly other intelligent species. On these issues there is a possibility of dialogue that will lead to a deeper understanding in both disciplines.

In this dialogue scientists will understandably play the role of reluctant prophets. After all the objective of science, and the proof of scientific theories, is the ability to predict the future for material phenomena - prophets of the material, not the spiritual. Yet their work can have religious implications as we have seen in our comparison of the development of the ancient concept of God as the Word and the new concept of the universe as a word.

16

Is God Evolving?

For many centuries philosophers and theologians have argued that God is perfect in Himself and therefore cannot change. If He changed then He would either have been imperfect before or would become imperfect afterwards. "You can't improve on perfection."

Modern science has raised another possibility. In many theories of various phenomena one encounters a situation where there are an infinite number of equivalent but different states. If this possibility occurs in simple physical theories is it not possible that there are many states of "perfection" and that God may be moving, or making transitions, between these states.

A practical example of many states of perfection is the set of all possible perfect mathematical circles. A circle is a type of curved smooth line with which we are all familiar. Circles can have any radius. There are an infinite number of circles - each with a different radius. Each circle is as perfect as any other circle. So we have an infinite number of perfect circles. God also may have an infinite number of possible states of perfection and he may be evolving between these states just as we can imagine a circle evolving by growing larger and larger.

God's transitions between states of perfection are not necessarily taking place in time which, after all, is a material artifact of man's consciousness and the universe of matter. The transitions between states of perfection can be viewed as the evolution of God.

What Differentiates the States of Perfection

If we think of the evolution of God between states of perfection the first question that occurs to us is what is the difference between these states?

Since God is a spiritual entity His primary distinguishing qualities that might be subject to change or evolution would appear to be His consciousness and His self-knowledge (which is an aspect of His consciousness). A number of theologians have suggested that the universe exists so that God may grow in the knowledge of Himself. Perhaps the universe is like a videotape extending from the beginning of time to the end of time that God looks at from "time" to "time".

Perhaps the states of perfection are distinguished by the stages in the growth of God's self-knowledge. Each state of self-knowledge is a state of perfection yet the states evolve towards a greater "end".

Why Should God Evolve?

Question: Why should there be an answer to this question that is logically acceptable to the human mind? Possibly God just evolves and the reason is not fathomable.

On the other hand, there might be a reason or rationale for God to evolve that we can understand. As stated above, an evolving God is consistent with theological ideas that hold the universe exists so that God may learn of Himself. "Learning is evolving." Perhaps God has an impulse towards self-knowledge.

The Evolution of the Universe

Our view of God as the evolving Word and our view of the universe as an evolving word are strikingly similar. The similarity evokes the image created by the theological notion that the universe is a reflection of the mind of God.

The similarity can be extended to a deeper level. The linguistic view of the universe is based on a quantum computer mechanism. The evolution of the universe is viewed as a quantum computer-like transition towards an unknown future. But since the evolution is quantum in nature it is inherently probabilistic. Chance plays a role in the process.

If the universe is an evolving computer "simulation" then one immediately thinks of Artificial Intelligence research and evolving computer programs that "get smarter" as they respond to events and stimuli. (In the area of computer science, developments in computer language and computers are driving to the goal of computer-based artificial intelligence.) It is possible that the universe is developing a form of Artificial Intelligence that gives it a knowledge of itself.

Perhaps the universe is moving towards a degree of awareness or self-awareness. We have seen man's intellectual growth over the millennia. It is likely that other

132

worlds have similar developments in progress - with all participating in a growth towards a universe of intelligence. (Perhaps project SETI, the search for extra-terrestrial intelligence, is a step in establishing a "neural" connection to another center of intelligence.)

Philosophers such as Pierre Teilhard de Chardin have speculated along these lines within an earthly framework. Teilhard de Chardin has suggested that the goal of evolution on earth is to eventually raise the world to a higher level of consciousness. The universe may also be evolving towards an eventual higher level of consciousness. This process may reflect the often-expressed view that the universe is a reflection of the mind of God. As God evolves, the universe may evolve in harmony.

The Word Evolves towards Self Awareness

With the linguistic view of the universe in mind, it appears plausible that God the Word is also evolving towards a larger degree of self-awareness. The ideas around the evolving universe tend to lend support to the theological view that God is evolving in His self-knowledge in conjunction with, and maybe in a way related to, the linguistic evolution of the universe. This thought raises the question whether the evolution of God impacts on the evolution of man.

Perhaps man's evolution to a deeper knowledge and a higher degree of morality parallels the evolution of God. Despite the unfortunate excesses of the Twentieth Century the average person in the more highly developed countries is measurably "better" than the average person in those countries in past centuries in terms of morality, sensitivity and education. When, in past centuries, did nations agonize over the death of one individual average person in combat or as a result of crime? Yet today the U. S. and apparently China (as well as other countries) will grieve over the loss of one soldier or pilot. Signs of a higher morality.

17

The Scientific Investigation of God

The preceding discussions of God: His evolution, His Nature, His self-awareness and so on raise a larger question: can we develop a science of God? Theology has been often described as the science of God. But in its current state it does not, for the most part, use the methods and thought processes of modern science. Many theological studies are experiential or personal; many theological works follow the Aristotelian logic approach of Thomas Aquinas.

135

We need to develop a new theology using the approaches and mindset of modern science - objective, analytical, and open to the input of scientific information.

In developing this new theology it appears reasonable to assume that God is open to rational, logical investigation - to assume that God does not embody internal logical contradictions.

Some might object that it is disrespectful to treat God as an object of scientific study. However the study of the Divine nature of God has been a continuing activity for thousands of years by theologians, philosophers, saints and mystics. Only the method is new.

A Scientific Model of God

Much is known about the nature of God: He is infinite, He is spirit not matter, He is outside of time but can intervene in human activities, and so on. These bits and pieces of information can be organized and developed into a "model" of God. A model is a conceptual design of an entity or theory that describes it in a logically coherent way. Scientists often create models of physical phenomena to help them understand and describe the phenomena.

The Science of Infinity

A scientific model or approach for the analysis of the nature of God might begin with the theory of transfinite (or infinite) numbers that was developed by Georg Cantor and others in the Nineteenth Century. Cantor was a great mathematician with a philosophical turn of mind. Although he was a mathematician he was a believer in a holistic view of nature and reality.

Cantor developed a theory of infinite numbers[26] that he called transfinite numbers. Cantor showed that there are many levels of numerical infinity. We normally think infinite is infinite.

Cantor developed methods to classify levels of infinity. We all know the number of positive integers is infinite. We all know the number of points in a line segment is infinite. Cantor showed that the infinity of positive integers is smaller than the infinity of points in a line segment. He then went on to develop a hierarchy of infinities. It is difficult to grasp that one infinity is more than another infinity. Yet this concept is true and can be mathematically proven (as Cantor did).

---

[26] See Georg Cantor, *Contributions to the Founding of the Theory of Transfinite Numbers* (Dover Publications, New York, 1955).

The Infinity of Infinities?

Amazingly there is no ultimate infinity. Cantor showed that the hierarchy of infinities grows without limit piling level upon level of infinity - an infinity of levels of infinity.

The Level of God's Infinity

On hearing of these discoveries we immediately wonder at what level of infinity is God[27]. We know that God has no spatial or temporal limits - He is space-time infinite. But He is also beyond space and time. At what level of infinity do these facts place Him? An open question. We do not know whether there are more dimensions then the ones that we are familiar with. Perhaps there are even an infinite number of dimensions. If extra dimensions are not an issue then God would have one of Cantor's infinities as His spatial-temporal infinity.

God is infinite in many ways: infinite Mercy, infinite Love and so on. The classification of these

---

[27] Cantor considered the notions of infinity of ancient and modern philosophers and also, notably, the Fathers of the Church. See page 55 and pages 73-74 of the preceding reference.

infinities within Cantor's hierarchy would be very difficult if not impossible.

We are tempted to place God at the ultimate infinity of infinities. But there is no proof.

Nevertheless, Cantor's concept of infinities is a step in developing a much deeper understanding of what we mean when we say God is infinite.

God as the Ultimate Time Traveler

God has many other qualities that could be analyzed as part of a scientific study of the Godhead. A fascinating quality is His existence outside of time coupled with His ability to enter time and influence the affairs of men. In modern terminology God is a time traveler - the ultimate time traveler. Perhaps the simplest view of God in this role is to imagine Him standing with a strip of videotape. The beginning of the strip is the beginning of the universe. The end of the strip is the end of the universe. God stands there as the videotape editor making changes here and there in the affairs of man and the universe. Perhaps sometimes He just watches - with a smile or a tear.

There are many paradoxes of time travel that have been explored scientifically in the scientific literature as well as in a literary way in science fiction. These studies can illumine the effect of God on history.

18

How Do We Discover Reality?

We have seen that science has developed theories of the universe that may describe its fundamental nature. These theories can be viewed as linguistic in form. They lead to a view of particles as letters in a cosmic alphabet and a view of the universe as a word. But how can we be sure that these theories reflect the reality of the universe? Are they perhaps mathematical fictions that are analogous to reality but are, in fact, not reality?

We are not endowed with an innate knowledge of the fundamental nature of the universe. We must learn of it.

141

We have been studying the universe for thousands of years. And no end is in sight. Are we on the right track?

To answer this question we must first answer another question: does the nature of human learning place limits on what we can discover about reality. Imagine a blind man trying to understand light. First, perhaps, he perceives it as warmth on his skin. Then through arduous investigation the blind man might discover light is an electromagnetic phenomenon. But he would never see a rainbow. His experience of light would always be limited. And his theory of light would be similarly limited.

Are Scientific Theories Subjective?

If we develop an understanding of the universe (a theory of the physical laws of the universe) is it subjective - colored by human patterns of thought and by human preconceptions? Or is it universal in the sense that a totally alien race would arrive at the same theory independently? We must examine the subjectivity of the learning process.

Consider an intelligent human being endowed with all the senses and possessing the ability to make instruments to discover phenomena not perceptible to the senses. How does that person translate observations and perceptions into knowledge of the reality behind the perceptions? For example, suppose we see a lighted candle placed in a closed jar. The candle soon goes out. How do we determine that the flame requires a substance in the air (oxygen)? Finding the explanation of a phenomenon is often not easy.

There seems to be three ways of coming to an understanding of the results of observations and experiments:

1. We may have an innate set of patterns of understanding. New perceptions and observations are understood within the framework of these innate ideas.

2. We may develop patterns of understanding from cultural conditioning or personal experience and then apply these ideas to observations.

3. We may have some innate patterns of understanding that are enlarged as the person matures through cultural conditioning or individual experience. This combined perspective leads to an understanding of observations.

Innate Patterns of Understanding

Surprisingly, some of our deepest thinkers have favored the existence of innate patterns of understanding. The most notable early example is Plato who developed a theory of learning in which learning is remembering the knowledge we had in a previous lifetime.

More recently, the psychoanalyst Carl Jung believed there are universal concepts and patterns of thinking in the human mind that guide our growth in understanding. He called these preexisting concepts *archetypes*. He believed these inherent concepts and ideas shaped our understanding of the phenomena of the external world.

The physicist, Wolfgang Pauli, also believed that the human mind had patterns of thought built into it that played an important role in our interpretation and understanding of natural phenomena. Pauli's belief is of special interest because of his deep knowledge of physics and his extremely critical nature. During his heyday from 1930 – 1955 he was known as the "conscience of Physics" because he authoritatively analyzed physical theories and pronounced them correct or incorrect.

The extent of his reputation is illustrated by a story based on his habit of sitting in the front row at physics lectures. If he liked what the lecturer was saying he would move his head up and down in agreement. If he disagreed with the lecturer he would shake his head from side to side. Many young lecturers had their confidence devastated by his disapproval.

From this behavior the story arose that Pauli died one day and went to heaven. He met St. Peter at the pearly gates who said, "God would like to have a few words with you." Pauli replied, "I would certainly like to meet Him and ask Him questions about how the universe was formed." Saint Peter took Pauli to God who said, "Please sit down." Pauli then said, "For some time I have wanted to ask You how the universe was formed." God then proceeded to explain how He created the universe. Pauli shook his head from side to side.

Pauli's penetrating intellect and his great contributions to physics (such as the Pauli Exclusion Principle for which he received the Nobel prize) lead us to give some weight to his belief that the human mind may have inherent patterns of thought that influence our interpretation of physical phenomena and the form of the physical laws we create to explain them.

Eastern vs. Western Modes of Thought

Recently some studies have appeared comparing eastern and western modes of thought. The studies purport to show that westerners tend to analyze phenomena and divide them into their component parts while eastern investigators (particularly Chinese investigators) tend to place phenomena within the context of larger systems seeking relationships within the larger system.

In fact western science embodies both modes of thinking. An example of the "Eastern" approach in western Science is Thermodynamics, which is concerned about the interrelationships between the components of systems "in the large". Thermodynamics seeks to deal with matter in bulk such as the pressure or temperature of gases and not to reduce the phenomena to the fundamental physics of atoms upon which Thermodynamics is based.

Western science is based on a very general mode of thought that includes modes of thought seen in non-western cultures. It appears that fundamental scientific theories are not subjective from a human perspective.

Fundamental Theory of Universal Significance

It also appears that fundamental scientific theories - particularly the Standard Model of elementary particles - are also not dependent on human thought processes. The reason is that they closely describe reality. Every type of particle has a corresponding representative in the Standard Model theory. This direct one-to-one correspondence at least suggests that the theory is perhaps the "simplest" possible theory.

A hypothetical alien might make a fundamental theory that differs from our theory but it would have to be more complex and convoluted. In addition it would have to be equivalent to our simpler theory. So a smart alien culture would develop similar equations and concepts although they would undoubtedly express them in a different notation (language).

19

Universality of Language

If the universe is linguistic - a word, and if God is the Word, then language becomes the most important material attribute of an intelligent creature.

Philosophers, linguists and others have long speculated on whether all human languages are based on a universal language. Some linguists make the strong claim that there was a common human language some 30,000 years ago.

Other linguists make a weaker claim to have found traces of a common basis for all human languages. Their evidence is not decisive and subject to dispute. But some major figures in linguistics and psychology have come to support this view. Jung and Minsky are notable modern advocates of the existence of a universal language.

A non-Species Specific Universal Language?

A more interesting question from our point of view is whether a universal language exists that transcends the human species. Is there some universal framework for communication that applies to the languages of all intelligent species? Can this universal framework for languages be extended to artificial languages and to the laws of nature? More importantly, are the languages of the Standard Model and SuperStrings within this framework?

Humans created the languages and theories that we have seen in previous chapters. To what extent is the description of Nature that we have assembled over the centuries universal and independent of human thought processes and modes of understanding?

The answer to these questions is very uncertain. St. Francis of Assisi and others have reportedly understood the language of animals. But modern man is unable to communicate with other species on this planet except in the most rudimentary way. We know that several species such as dolphins, porpoises, whales, monkeys, apes and elephants have some intelligence.

But we are not able to communicate with any of them in their own language. One aspect of the problem is the "language" of some of these species is partly non-verbal in the sense of not being based exclusively on sounds. On the other hand, aren't these differences precisely the things that would help answer questions about a universal language?

Other Intelligent Species on Earth

The preceding questions are both philosophical and practical. We are slowly beginning to realize that some animals are at least semi-intelligent. Many primitive peoples that are closer to nature than civilized man believe animals have "souls", personalities and understanding. For example, the American Indian would often apologize and thank an animal after killing it for food.

Modern man is slowly coming to realize that the place to look for intelligent life is not only outer space but also on the earth. To that end the first conference on intelligence in earth species took place in the Fall of the year 2000. This conference brought human students of various semi-intelligent species: dolphins, elephants and so on to compare notes. Perhaps the most amazing thing about this conference is that it is the first conference where specialists in different animal species convened together.

The continuing stream of discoveries on animal intelligence and language abilities leads one to wonder if we are on the Planet of the Humans unable to appreciate the communication skills and intelligence of our neighboring species.

Communicating with Other Earth Species: Dolphins

Consider the case of dolphins. Dolphins are acknowledged to be intelligent and verbal. They are masters of using sound to communicate. After perhaps twenty to thirty years of intensive study of dolphin communications we are unable to communicate with them in their own language. Rather we try to develop rudimentary forms of pseudo-language that we can teach them.

There have been reports that dolphins have role-reversed with their human investigators and tried, on occasion, to find the range of frequencies that humans might be able to understand. They have run through various ranges of sound frequencies trying to find a match with their dyslexic human investigators. Which species is the intelligent one?

Recently[28] a researcher in dolphins has found the following remarkable traits:

1. Dolphins learn and can repeat intricate signals from their friends – an important part of language skills.
2. Dolphins appear to be capable of vocal learning – a prerequisite for language skills.
3. The signaling pattern of dolphins is similar to that of pre-historic humans when they first developed organized speech according to experts.
4. Dolphins use matching whistle patterns to address each other. These patterns seem to signal membership in a social group.
5. Dolphins adopt a unique signature whistle pattern when they are young. The whistle pattern is used like a name. A dolphin calls to another dolphin using its name (whistle pattern). Dolphins keep their name as they grow older.

---

[28] Vincent M. Janik, Science, August 25, 2000 issue.

Given these scientifically proven facts and what remains to be uncovered can we deny their intelligence and language abilities – even if they are quite different from ours?

Can we learn to communicate with dolphins by electronically generating sounds at frequencies they use and then begin communication by finding the names that they give objects as we do with young children. Can we then learn how they combine words in "dolphin grammar"? Should we not also learn to communicate with other semi-intelligent species such as apes and elephants?

Our profound lack of understanding of communication with earth's other species with which we share a common environment and with whom we can meet "face to face" raises the question: Would we recognize communications from an extraterrestrial species if we received them? And how would we interpret them?

Project SETI has great and noble goals of establishing contact with other civilizations by detecting their transmissions. Can it succeed? Could it recognize those transmissions if received? What Rosetta Stone would guide us to understand the content of these communications? We can't read transmissions from earth's past – vast amounts of writings from ancient civilizations. How can we decode and translate electronic messages with

154

no common framework for understanding? In both cases we are dealing with one way communications – from the past, and from a great distance (and also in the past because of the long time it takes for signals to reach us from the stars).

Can We Communicate with Alien Species: Project SETI?

Project SETI might be well advised to start by understanding transmissions from intelligent species on earth before listening for alien communications from beyond the earth. An interesting test for SETI would be to take a conversation between two dolphins, "pipe" them to its receiving station at a random time, and see if the non-random dolphin communications would be uncovered by its detection process or regarded as random noise. After all, if our civilization is any guide, most communication is for entertainment and conversation.

While this is not the place to critique project SETI, with its admirable purpose, a significant issue that SETI must face has been raised by the development, and increasingly widespread use, of encrypted data transmission. An important recent development in encryption is to embed encrypted messages in pictures or other large chunks of innocuous data. Would not other civilizations send long range messages of importance in an encrypted format for the same reasons of security that we have on earth? Could messages be embedded in "static" using advanced encryption techniques that we would have no chance of deciphering without a knowledge of the encryption algorithms and encryption keys? If this is so, then we will only be likely to detect civilizations at a similar level as our own in our immediate galactic neighborhood. This possibility appears to substantially lower SETI's probability for success.

Our best hope is to catch transmissions from civilizations in the stars while they are in the happy interlude corresponding to our civilization between the Roaring Twenties (the practical beginning of radio) and 2001!

Languages for Physical Theories

The development of a language or languages for physical theories is in a way related to the question of the languages of human and other species. We express our thoughts with language and language in turn helps shape our thoughts. The direction of physical theory, particularly the development of physics after the Standard Model, is in part conditioned by our way of expressing current theory.

In elementary particle physics we use the language of Quantum Field Theory because it is where we have been led by experiment, and by our attempts to understand experimental results. There is an element of truth in earlier physical theories and looking back we see these earlier theories as approximations to the current theory. This continuity is both comforting and necessary – the scientists of past centuries developed profound and interesting theories at the level at which they were working.

But one cannot help wondering if the path that the development of physics followed is the only path. This question has been expressed in a dramatic way by Edward Witten, a leading String Theorist. Mr. Witten suggests we were doing 21st Century physics in the Twentieth Century due to the "accidental" discovery of the beginnings of String Theory in the early 1970's by Gabriel Veneziano and Mahiko Suzuki.

Physics on Other Worlds

Can we imagine dolphin-like creatures on a watery world discovering fire, developing alchemy, finding the theory of the wandering worlds of their solar system, and having a Newton-dolphin develop a theory of gravity in a watery setting? Is it easy to imagine further that they developed a technological civilization that provided the underpinnings of a theoretical development leading to the Standard Model or SuperString theory? It seems that our ability to develop in Science and Technology was based on very fortunate features of earth – a thin atmosphere, abundant metals, abundant fuel, and solid ground.

It also appears that physics would be much harder to develop on a watery world. And the progression to our type of technology on a watery world is hard to imagine. For example, instead of forging metal in a furnace, metal objects on a watery world might be made by slow chemical deposition techniques.

So, at first glance, the overall framework of the development of physics would appear to be "easy" only on an earth-like world although the composition of the atmosphere and other details could be quite different.

If this is true, then can there be other descriptions of the universe based on other physical languages that perhaps are equivalent to the language of the Standard Model but differ greatly in form?

Is There a Best Formulation of a Physics Language?

If there are equivalent, and yet very different, formulations of the physics of the Standard Model, then is there a "best" formulation? In computer languages, and in physical theories, there is often a preferred language to perform calculations or to define a theory. Is the Standard Model the best way of representing the physics of elementary particles? At the moment the answer is yes because there are no other known equivalent formulations. The language of Quantum Field Theory is the only apparent way to express our deepest physical understanding of real world experiments.

Although SuperString Theory has many adherents it has not been shown to account for the features of the universe as we know it. It may succeed in the future.

In the past there were choices between equivalent theories. In the case of planetary motion the Ptolemaic Theory was initially more accurate then the Copernican Theory but the Copernican theory was simpler for performing calculations.

The Copernican Theory was also a better description of reality and led eventually to the Newtonian theory of gravity. It would have been extremely difficult to develop the theory of gravity from the Ptolemaic Theory.

The Copernican Theory was clearly the preferred description of planetary motion. It provided a path to the development of Newton's theory of gravitation.

Another situation where there were competing approaches was in the development of Calculus. Newton developed one notation (symbolism) for calculus and Leibnitz developed a different notation. The notation of Newton was cumbersome and not as easy to use, or as intuitive, as the notation of Leibnitz.

As a result continental European physicists made significantly more theoretical contributions in the Seventeenth and Eighteenth Centuries than English physicists who were constrained by the Newtonian notation.

After English physicists adopted the Leibnitz notation in the early Nineteenth Century, English theoretical physics took off as we can see in the work of Maxwell, Kelvin and Rayleigh among others. Conclusion: the notation that we use in Physics can significantly help or hinder the progress of Physics. The Leibnitz notation was clearly superior.

A third example of equivalent theories appears in Quantum Theory. Heisenberg developed a theory of quantum mechanics called Matrix Mechanics. Schrodinger developed a theory of Quantum Mechanics called Wave

160

Mechanics that was based on a wave picture of quantum effects. While the theories were completely equivalent, Wave Mechanics led to Quantum Field Theory and thus it was the theory that led directly to the next stage of Physics.

We conclude that theories may be equivalent but often one theory is preferred because it provides a path to the future.

Currently, there is only one theory – the Standard Model – that describes Nature as we know it. There is no successful alternate model. (The SuperString Theories are at a deeper level and have not been shown to successfully describe Nature despite thirty years of work.) We have seen that the Standard Model and SuperString Theory can be viewed as computer languages. Computer languages normally have many possible equivalent languages. This fact suggests that equivalent formulations of the Standard Model may exist. However, due to its close relation to the particles and interactions observed in Nature the Standard Model may well be the simplest formulation and may be the path to the future. Although the simplest formulation was often the path to the future in recent physics theories we must remember that there is no guarantee that the simplest formulation will always be the path to the future.

The Standard Model is a highly mathematical theory with many abstract features such as complex symmetry groups. Could we have done without all this mathematical baggage and developed a different view of the universe – perhaps like holistic dolphins on a watery world? It appears that the answer is no.

The holistic view of Nature is essentially qualitative and descriptive. It does not lead to successful progress in science. For example, the ideas behind quantum mechanics were generated from numerical discrepancies from the predictions of classical theory. Numbers are the key to the universe as Pythagoras said many centuries ago.

On the other hand we cannot dismiss holistic theories completely because they appear to have an element of truth in certain situations. One wonders if some ideas of holistic theories are guidance from the eternal. Sometimes numerical, physical theories are developed that have an uncanny resemblance to holistic ideas. The concept of the universe as a word is an example.

Why is Mathematics Required in Physical Theories?

The mathematical formulation of physical theories is an absolute essential. The reasons for a mathematical formulation are simple. The following list gives the rationale for the role of mathematics in physics:

162

- Physics measures quantities – Numbers are needed
- Physics is concerned with rate of change – derivatives are needed
- Physics must calculate numbers – combinatorial mathematics is needed
- Physics is concerned with shapes – topology and geometry is needed

The human in us hopes the mathematics of physics can be re-expressed in simple terms that we can understand. Feynman said any good idea in physics should be expressible in a few simple words. Many parts of physical theories can be expressed in simple words. Many parts are not easy to express in simple words.

The current major problems in Physics are "What is the next step?" and "What does it all mean?" The answer to the first question may be SuperString Theory. The answer to the second question may be another question raised by Lord Byron, "Who will then explain the explanation?"

Language appears to be the key to the understanding of the universe. We have been driven by experiment to a language that is strange and unexpected. The language of the Standard Model cannot be justified by more fundamental principles or concepts at this point in history. The next stage of physical theory can be expected to be equally strange, unexpected and exciting.

20

What is Reality?

If material particles are letters and the universe is a word then we have come to a view of Nature where the substantial things of everyday experience have dissolved into empty space. In fact space and time themselves have disappeared. We are left with a universe consisting of structure imposed on the "formless void." (Genesis?)

Events are uncertain and described by the probabilistic ideas of Quantum Theory. The language of the universe describes an insubstantial reality – nothing – a mathematical structure without substance.

The surprise is that it took centuries of scientific research to reach the dissolution into the insubstantial.

165

Religious thought reached that conclusion a long time ago - not only in the West but also in Buddhism.

The insubstantial world of the Standard Model with particles that are tiny abstractions in a mostly empty space leads us to inquire about the Nature of Reality itself. The even more insubstantial world of SuperStrings where nothing remains except mathematical laws and mathematical SuperStrings makes the inquiry crucial.

While it would be nice to think of the Platonic World of Ideas as the essence of Reality somehow it does not capture the reality that is emerging from our continuing encounter with Nature. The chaotic, probabilistic world of quantum reality is certainly not the serene world of Platonic Ideas.

Dream-like Reality

The world that we have uncovered seems more like an evolving dream or experience of consciousness with a certain logic and permanence, but tinged with an edge of uncertainty and a perverse absurdity at its fringes.

Dreams are the only things of common experience that are close to our current concept of the universe – substantial but insubstantial, logical but illogical, built out of thoughts yet seemingly real. One is tempted to ask – are we part of a dream?

> Reality?
> What we see.
> What we don't see, that's really there.
> What we don't see that influences what we do see.
> Theories that lead to theories of greater emptiness
> Until Reality adopts the substance of a Dream.

Theories of Everything

Impressions are easy to form and are as often wrong as right. Our discussions have led us to a deeper and deeper view of the nature of our universe. It is as strange and wonderful a place as one could imagine. And in fact it passes imagination.

No one foresaw the Theory of Relativity or Quantum Mechanics. Nature led us to these unimagined theories kicking and braying along the way. Current popular fundamental theories that some have dubbed "The Final Theory" or "The Theory of Everything" are more likely to be another stage in our encounter with Nature and Nature's God.

Just as people have talked of "The End of History" as if there would be no more wars or conflicts, or Stalin's or Napoleon's or Caesar's – physicists now talk of an "End of Physics" where all problems will have been solved (at least in principle) and our understanding of Nature will be complete.

A similar feeling – in both the political and scientific spheres –appeared around 1900 and at the peak of the Roman Empire. In both cases unforeseen events shook the worlds of Politics and the Sciences.

The final chapters in these sagas remain to be written. We are not, after all, anywhere near the "last syllable of recorded time".

Cosmos and Consciousness

The fundamental theories of space, time and matter have led us to a view of Nature as composed of insubstantial things – mere nothingness – yet extraordinarily complex in form, and yet again having elements of simplicity. The more we understand of Nature the more we are lead to the impression of a "nothingness" described by a complex language. In some sense there is an analogy to human thought - consciousness. We think, we dream. Yet there is no substance to these thoughts. And yet there is a complex underlying set of rules and chemical laws that somehow are at the root of these thoughts.

Wolfgang Pauli – one of the severest critics of Physics in his time and a practical, "level-headed" person – wrote essays[29] suggesting that there was a notable resemblance between our consciousness and the idea of a physical field such as an electromagnetic field. He referred to comments of William James[30] comparing a magnetic field to human consciousness with its "whole past store of memories [that] floats beyond its margin... the entire mass of residual powers, impulses, and knowledge that constitute our empirical self ... So vaguely drawn are the outlines between what is actual and what is only potential at any moment of our conscious life." A magnetic field – or for that matter the quantum field of an elementary particle – "floats" in space and time and contains within itself a store of data. The field can influence other particles just as our consciousness can influence our actions.

This analogy of the insubstantiality of the conscious mind with the insubstantiality at the root of Nature leads

---

[29] Wolfgang Pauli, *Writings on Physics and Philosophy* (Springer-Verlag, Berlin, 1994).

[30] William James, *The Varieties of Religious Experience*, (Modern Library, New York, 1902) pp. 226, 227.

one to consider the view that Nature is at its core a thought (or set of thoughts). The universe might be the nothingness residing in the consciousness of an Observer. Are we "such stuff as dreams are made on"?

How do we know we are not part of a greater being's consciousness or dream? More especially, suppose dream events take place in a logical ordinary way. Is there a test for Reality – whatever that means? Can we distinguish a logical Dream from Reality? The Buddhists say no. Yet humans can develop the ability to distinguish a dream from reality while within the dream.

The new view of the fundamental theories of nature as definitions of languages is one step closer to Wolfgang Pauli's inspired hope of a universal science encompassing both mind and matter.

Divine Dream - Not Buddhist Reality

The idea that the universe is a nothingness and that it has the flavor of a dream is similar to Buddhist thought. But the Buddhist views consciousness as the creator of reality and the shaper of the form of the universe. Mind creates the universe and gives it form.[31] All conscious entities participate in the continuing creation and evolution of the universe. In some sense they are part of a universal consciousness. Buddhism does not have a God in an absolute sense although it does have demons and deities that are in part created by the mind of the believer.

Our view is that God defines reality and shapes the universe. The dream-like quality of the basic nature of the universe reflects God's design of the universe - not individual consciousness to any significant degree. So speculations that man can influence or determine reality solely through thought are not implied by our discussion. They are also not ruled out by our discussion.

The question of whether human consciousness can influence reality directly is a scientific question that has not been resolved.

---

[31] See, for example, John Blofeld, *The Tantric Mysticism of Tibet*, (Arkana, Viking Penguin, New York, 1992).

21

Human Consciousness

A View of Consciousness

There are numerous views of Consciousness. Some of these views attempt to make distinctions between consciousness, the mind, and the brain (body). The mind is a nebulous thing we associate with consciousness, feeling and thought. The body, and in particular the brain, is obviously connected to the mind and supports the mind's activity. Yet Consciousness seems endowed with miraculous abilities that many find hard to base entirely on the properties of the brain.

These considerations have been raised to the level of a problem: the Mind – Body Problem. How can the mind and its glorious properties be based on the morass of cauliflower-like flesh that we call the brain? One simple and seemingly facetious resolution of this problem is suggested by the discussions of the preceding sections: the universe is nothing, there are no bodies, thus there is no problem – the brain is an illusion. No body, no Mind-Body Problem!

There is an element of truth in this facile solution. In the preceding chapters we saw how the path of physical theories led us to a view of reality as nothingness given structure by physical law. We also saw how our view of matter was an illusion based primarily on electromagnetic forces. So the human brain is in a sense an electromagnetic illusion. *The brain is just as insubstantial as the mind in reality.*

The difference is that our senses can perceive the body but cannot perceive the mind and its contents in a direct way. We are influenced by our senses to attribute a greater reality to the body. Perhaps we are influenced by our senses similarly to attribute less reality to the spiritual.

174

The Power of Simulation

There is a deeper sense in which the Mind-Body Problem is resolvable. We generally do not appreciate the power of electromagnetic circuits to create illusions. We see the brain as a hodgepodge of electromagnetic circuitry based on neurons and other brain structures. We then view the mind, and its unity, clarity, powers of logic and analysis, and other features that compose one great entity. What a difference!

It is difficult to reconcile the unity of consciousness of the mind with the brain that implements it. Yet it is more difficult to deny that the mind is based entirely on the brain. Modern research[32] clearly shows the dependence of the properties of the mind on the features of the brain. Consider the effect on the mind of brain diseases or of injuries to the brain. Consider also recent work that has associated the right frontal lobe of the brain with a person's sense of self. A researcher has shown that patients who have brain damage within that area of the brain lose their sense of individual selfhood. (The sense of self is a very special

---

[32] Gerald M. Edelman and Giulio Tononi, *A Universe of Consciousness*, (Basic Books, New York, 2000). There are many other excellent books on consciousness. See the references in Edelman and Tononi or search the Web.

feature of man. It appears to be missing in most animals except for dolphins and a few of the primates.)

Modern computer technology actually offers a very clear analogy to the relationship between Consciousness and the brain. Consider a modern Personal Computer, a PC. If we open it up we see an ugly hodgepodge of chips and computer circuitry. By only looking at the innards of the PC we have no concept of what this electronic menagerie can generate.

Then we turn on the PC and see the fabulous graphics of a modern computer operating system: lots of windows containing exciting graphics and multimedia displays. We can manipulate these windows causing them to change, disappear, reappear with new content, and so on using a mouse, the keyboard or a joystick. We can run captivating multimedia games and simulations with the click of a mouse or the movement of a joystick. We can access and manipulate external information from around the world using the Internet.

Does the computer screen look in any way like the innards of the computer? Does the unity, sophistication and flexibility of the display reflect the odd collection of electronics inside the computer? Obviously not.

Figure. The ubiquitous PC. From low level electronics we obtain intricate, organized windows displays.

This example is directly analogous to the relation of Consciousness and the brain. The thoughts, unity and activity of Consciousness (the "screen") have no obvious connection to the details of brain activity (the "computer innards"). Yet the mind is a construct of the electrical activity in the brain.

The Consciousness of the mind is the combined result of the electrical activity of the brain.

The spiritual part of man is another question.

177

A Probabilistic Computer Model of Consciousness

Although human Consciousness is large and complex it must be finite since it is derived from the human brain which is finite although it contains an incredible number of parts.

Since human Consciousness is based on a finite human brain, a sufficiently large and complex computer can in principle, simulate it. Therefore it seems reasonable to think that Consciousness can be modeled on a computer with appropriate features and capabilities.

Does Consciousness "run" like a computer program? No. A better concept of Consciousness is to view it as a set of parts with capabilities and features that interconnect to constitute Consciousness. Each of these parts may map to groups of neurons in the brain.

One can think of each part as an "object" that has a specific set of capabilities and features. These objects have a "mini-program" inside them that specifies their behavior, and how they hook up with other objects to perform tasks that are aspects of Consciousness. The hooks are variable and dynamic.

The time evolution of the Consciousness from state to state is a result of the execution of these "mini-programs" in a dynamic ever-changing way. There is no overall program but instead there is an ever changing, dynamic, unfolding of states of Consciousness that respond to external inputs based on the previous state of Consciousness plus random events within Consciousness.

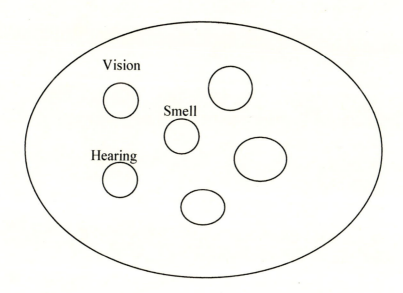

Figure. An Object-Oriented view of Consciousness.

The description of Consciousness as a collection of objects with features, and with an internal "mini-program" describing the object's behavior and interaction with other objects, can be called an Object-Oriented definition. Objects such as these are part of Object-Oriented Programming - the preferred way to program among computer software developers.

The Object-Oriented model of Consciousness seems to be well adapted to many of the observed features of Consciousness. For more details see the author's book *Cosmos and Consciousness*.

Cosmos Computer vs. Consciousness Computer

We have developed a linguistic Computer Theory of the Cosmos. We have developed the beginnings of a computer theory of Consciousness.

We have arrived at a unified view of the Cosmos and Consciousness. Consciousness and Cosmos are in a way "the very images of each other". The universe, and man's consciousness, reflect the mind of God.

INDEX

181

Copernican, 159

Copernican theory, *159*

cosmic alphabet, 50, 82, 141

Cosmos, 168, 180

Cosmos and Consciousness, xvi

creation of particles, 55, 66

cultural conditioning, 143

D

dark matter, 54

Darwin, 124

data, 54, 64, 65, 66, 67, 71, 155, 169

Democritus, 91

dolphins, 151, 152, 154, 155, 162

Dreams, 166

E

ecstatic, 116

Egypt, 3, 4, 5, 6, 9, 11, 15, 16, 19, 37, 39, 40

Einstein, 60, 99

Eleazer of Worms, 22

electromagnetism, 52

electrons, xvi, 48, 49, 50, 52, 55, 61, 62, 68, 76, 77, 79

elementary particles, 55, 57, 69, 159

encryption, 155

Eternal, 34, 36, 167

Eternity, 34

evolution, 123, 125, 128, 130, 132, 133, 134, 135, 171, 179

Exclusion Principle, 145

Ezechiel, 93

F

Faustus, 37

fermion, 54

Feynman, 100

Feynman, Richard, 61, *163*

Final Theory, 167

Finnegan's Wake, 48

flying saucers, 90

fringe science, 122

G

gematria, 22, 23, 25

Generality, *142*

Lutherans, 37

M

magic, 11, 12, 16, 22, 37, 42

Marie of the Incarnation, 36

mathematics, *162, 163*

Matrix Mechanics, 160

Maxwell, James, 160

Melchisedech, 17

Mephistofeles, 37

Mesopotamia, 7

Middle East, 7, 32

Mind – Body Problem, 174

mini-program, 180

Minsky, Marvin, 150

mystical, 23, 24, 33, 36

N

Name of God, 26, 32

Nature, **150, 161, 162, 165, 166, 167, 168, 169**

NDE, 103, 104, 105, 107, 111, 112, 118, 119

Near Death Experiences, 103

neurons, 175, 178

NeutrinoMan, 174

neutrons, 48, 49, 76

New Kingdom, 11

Newton, *158, 160*

nothingness, 168, 170, 174

nucleus, 48, 49

O

Object-Oriented Programming, 180

observations, 142

Origen, 33

Osiris, 5

out of body experiences, 104, 114, 119

output, 68

P

patterns of thought, 142, 144, 145

patterns of understanding, 143, 144

Pauli, Wolfgang, 144, 145, 169

perceptions, 142

perfection, 129, 130, 131

Personal Computer, 176

photons, 50, 60, 61, 62, 68

physical laws, 142, 145

Plato, 4, 20, 144, 166

Platonic World of Ideas, 166

polytheism, 5

positrons, 61, 62

Predestination, 101

Primeval Mound, 27, 28

Probabilistic Computer, 178, 180

production rules, 70

program, 178, 179, 180

Project SETI, 154, 155

protons, 48, 49, 56, 76

Ptolemaic, 157, 159

Ptolemaic Theory, *157*, 159

Pythagoras, 4, 162

Q

Quantum Field Theory, 157, 159, 161

Quantum Mechanics, 167

Quantum Theory, 160, 165

quarks, xvi, 48, 49, 50, 52, 55, 68, 76, 79

R

Ra, 11, 14

Rahner, Karl, 36

Rê, 6

Reality, 23, 24, 32, 33, 45, 135, 141, 142, 159, 165, 166, 167, 170

Relativity, 167

Rosetta stone, 154

S

Saint Peter, 145

Scheffler, Johannes, 35

Scholem, Gershom, 23, 24

SETI, 133, 155

shaman, 119

Sitwell, Sacheverell, 87

St. Francis of Assisi, 150

St. John, xvi, 32, 33

Standard Model, 51, 52, 53, 54, 55, 56, 60, 62, 63, 67, 68, 69, 70, 75, 78, 79, 150, 157, 158, 159, 161, 163, 165, 166, 170

string, 55, 57, 73

strong interactions, 52

185

About Stephen Blaha

Stephen Blaha is an internationally known physicist with extensive interests in Science, the Arts and Technology. He has been an Associate of the Harvard University Physics Department for over ten years. In addition he has written a highly regarded book on physics, consciousness and philosophy - Cosmos and Consciousness, and books on Java and C++ programming. He has been on the faculties of Cornell University, Yale University, Syracuse University and Williams College. He was also a Member of the Technical Staff at Bell Laboratories. Dr. Blaha is noted for contributions to elementary particle theory and solid state physics theory as well as contributions to Computer Science. Dr. Blaha writes with a clear, lucid style that makes his books a joy to read if only because they are short, to the point and have a high density of intriguing ideas.

Printed in the United States
2474

9 780759 663046